An Introduction to
Muscle Powered
ultra-light gas
BLIMPS

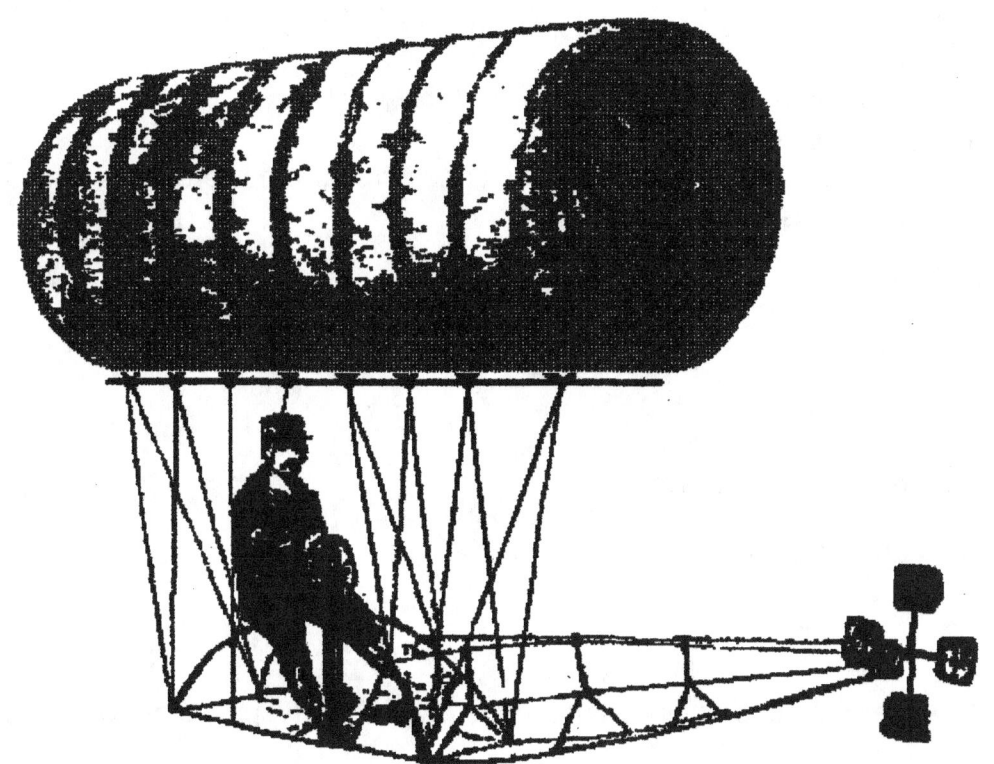

In All Their Glory

Assembled by
Robert J. Recks

MUSCLE POWERED BLIMPS

Copyright Robert J. Rechs, 2002

All rights to bound volumes of this book are reserved for the author, who is wholly responsible for its content.

Inquiries should be addressed to the individual author (r.recks@juno.com) or to the publisher (piolenc@archivale.com).

Association of Balloon and Airship Constructors (ABAC)
P.O. Box 3841
City of Industry, California 91744-9991
http://abac.archivale.com
catalog: http://archivale.com/catalog

Library of Congress Cataloging in Publication data:

Muscle Powered Blimps
Bibliography: p. no index
Summary: An introduction to what has been done in muscle powered gas blimps used for sport and research, as adult reading.

1. Aircraft
1. Airships
1. Blimps
1. Author

ISBN 0-937568-30-9 (hardcover edition)

ISBN 978-1540383709 (paperback edition)

TL685.1 C++ 629.133

To order/reorder the paperback edition: https://www.createspace.com/6715592

CONTENTS

Foreword... 1
Introduction... 2
Acknowledgements... 2
Ultralights.. 3
Chapter 1: History... 5
Chapter 2: Design Considerations................................ 35
    Records.. 44
Chapter 3: Details and Compromises.............................. 47
    Notes to FAI Sporting Code............................... 49
    Airship Mathematics—predicting speed..................... 51
    Cycle Power Chart.. 52
    Body Positions... 53
    Gondolas... 54
    Working with Aircraft Tubing............................. 55
    Power Transmission....................................... 57
    Propellers... 62
    Envelopes.. 75
        Ultra Light Fabrics.................................. 80
    Retractable Grappling Hook............................... 85
Chapter 4: Ergonomics—Getting Fit, Keeping Fit.................. 87
Chapter 5: Rocket Science for Pedal Power....................... 97
    Propeller Theory... 98
        Lifting-Line Formulation for Counter-Rotating Rotors.... 105
    Propeller Airfoil Coordinates........................... 114
    The Aerodynamics of Human-Powered Land Vehicles......... 118
    The Screw Propeller..................................... 128
The White Dwarf Man-Powered Blimp.............................. 140
    Man Powered Blimp....................................... 149
    The Flight of the Brazilian Dirigible Caloi............. 153
    White Dwarfs and Pedal Pushers.......................... 157
    Who is Bryan Allen?..................................... 163
    Building and Flying a Human-Powered Airship—White Dwarf. 168

The Brazilian Dirigible Caloi 179
Vintage Designs
      Building Airships (by Glenn H. Curtiss, 1910) 186
      Letter from C.A. Nichols, Jr., 1 June 1921. 189
Shimano Bicycle Parts Catalog 190
      Glossary 192
      Gears 199
Contacts 210
Bibliography 211
Related Books 212
Organizations 213
Conclusion 214
About the Author 215

FOREWORD

So you let the *Snoopy* blimp run away with your imagination? Let me be the last one to discourage you. It can be done, but NOT without LOTS of money, and a lot of blood, sweat, and tears. Yes, they do look simple; and they are, basically. But be careful; once it is inflated, your **personal safety is on the line.**

This book was assembled from my personal archives, at the request of several purchasers of "Building Small Gas Blimps", much to my dismay. I had no idea that so many enthusiasts of Radio Controlled (R/C) models were dreaming of that "Ultimate." The Small Gas Blimp **book** is still the best "how to do" book applicable, but this one will be **expanded as questions arise.**

This book is NOT intended to be a design manual, but rather a guide as to the successes in muscle powered blimps. At this point, you no doubt have some new ideas to test, and dreams to develop. I say again, do not let me discourage you. But you do owe it to yourself, and the sport airship community, to visit several successful airship operations. See for yourself what it takes to handle a blimp in high winds & no winds, and in temperature extremes. Talk to the crews, kick the tires, feel the fabric, and by all means, **ask questions.**

The most important thing I want to get across to you is safety. Not just to the people you fly over, but YOU in particular. You won't find much in here on loads, stresses and yields of materials. You, as the pilot are going to have to set the limits. Even the Wright Brothers went through the school of hard knocks (Wilbur was almost killed). Expert advice is available; at least listen, and give some thought to all suggestions. **Do not compromise safety,** even if it costs a little more money.

Lastly, the designs and techniques in this book are NOT my ideas. They have been tried and tested over the last 100+ years. The newest innovations in the last 50 years are the composite technology readily available, The best for muscle-power, of course, is the development of *"high*efficiency"* low-speed propellers. **But I do not warrant anything in this book for your particular application. So** please do not call me at 2 in the morning (from your hospital bed?) to ask me what went wrong. O r at anytime if you call yourself an *"airship* manufacturing company".

The author
(30 *years experience building & flying blimps)*

* For setting records and winning races.

INTRODUCTION

This endeavor is the result of many requests from the Internet for information on Pedal Powered Blimps. I chose to call this "***Muscle-Powered*** Blimps" to better show how it got here historically. And it just so happens, that a friend of mine (Don Dwiggins, now deceased), has already reserved that title.

I would like to think of this as an information "source" book. It is definitely *NOT* intended to tell you how to build or fly one. But rather to show you what has been done, and give you some guidance on the state of the art. There are many good airship related organizations with members involved in every aspect of this unique aircraft. Please, join at least one of those listed in the "Sources" chapter.

This is not my first experience in compiling a technical book, nor will it be the last. It ***IS*** to be considered as a companion volume to my "***Building Small Gas Blimps***" in publication since 1977 (revised 1987 & 97). I certainly hope that you obtain a copy, and read it from ***front*** *to back*. It contains much more detailed technical information critical to the safe construction & operation of the finished product.

My experience also includes more than 35 years with blimps, as an engineer, production & field mechanic, pilot-instructor, and FAA Airworthiness Representative (DAR). In short, aviation has been my lives work, blimps my first love.

ACKNOWLEDGMENTS

I want to take this opportunity to thank the many friends and organizations that have supplied information over the years, including: Brian Allen, JPL Engineer; Tracy Barnes, Pres. The Blimp Works; Eric Brothers, LTAS Pres.; Tom Crouch, NASM Curator; Arnold Naylor, RAeS Curator; Mark Peaslee, Blimp Designer; Carl Stefan, LTA-DER; Jim Winker, RAVEN Ind. Pres. ret.; and my many fellow members of the AIAA Lighter than Air Technical Committee.

The book reference sources are listed in the Bibliography. But I would like to personally thank the Authors of the articles in the Appendices, and hope that they are given their due allocates for their research and technical writing.

ULTRALIGHTS

Background: Just to clarify an important point, the Federal Aviation Administration (FAA) gives the word very specific limitations. If you stay within the guidelines, it can save you some nickels and dimes, but you will still have specific limitations on what you can do and where you can operate. A home builder can NOT claim ignorance if challenged. The complete rules (Federal Aviation Regulation, FAR-103) are available from the Govt. Printing Office, or your local FAA Flight Standards Office. And I suggest you acquire a copy AND keep it available.

Definitions: The specific rules include an aircraft empty weight (E/W) not exceeding 254 pounds. To a blimp, this means the entire aircraft BEFORE any lifting gas is added. (There are some exceptions for "safety" equipment) The rules also stipulate that this category was designed for the "homebuilder/pilot" as the only "owner/operator" of the aircraft. Yes, some ultralight airplane builders you have seen do push (over) the limits, but one challenge will initiate an investigation. The gray area comes when you put in an extra seat for a passenger and call it "student training". The other gray area is when you see one flying over your house, obviously out of the designated ultralight flying area.

Benefits: For you as a blimp owner, **NOTHING**. Yes, you eliminate 30 minutes of paperwork and a few dollars. But you are **NOT** required to have a FAA Pilot License to fly your *blimp* regardless. Even if you do not qualify as ultralight, you can still apply for a "Student License" (free) and request a "waiver" to fly your "experimental" aircraft in a mutually agreed upon "flying area". Be sure to mark an X in the block "crew training" of your FAA application Form 8010-6.

Is it Possible to build a blimp with an E/W of less than 254 pounds? Certainly, and I don't mean out of *Saran-Wrap*. I have seen TWO good 2-seaters under 9000 cu.ft. with single Konig engines that worked very well. It is easy to strap a small engine to a lawn chair for a gondola, with a 6 mil. heat sealed *Polyurethane* envelope. One of the two above had the engine gimbaled for directional control, the other had single layer tensioned fins. But then, Santos Dumont & the Wright Brothers had pretty flimsy designs also.

Some Numbers: One *Airship Homebuilder* is presently contemplating the construction of a blimp with the following target weights:

Envelope:	63 #	9000 cu.ft. (3 oz/sy)		
Ballonet:	21 #	10 %	Tail Fins:	35 # 115 sq.ft.
Propulsion	50 #	15 hp.	Framing:	40 # Alum.tubing
Cabin:	15 #	enclosure	Nose:	20 # structure

Total equals 244 pounds, leaving only 10 pounds for valves, ducts, control cables, instruments, and comfort (?). I might add that the pilot weighs about 140 # wringing wet.

It CAN be done,
but it is NOT something that I feel comfortable flying outside the hangar.

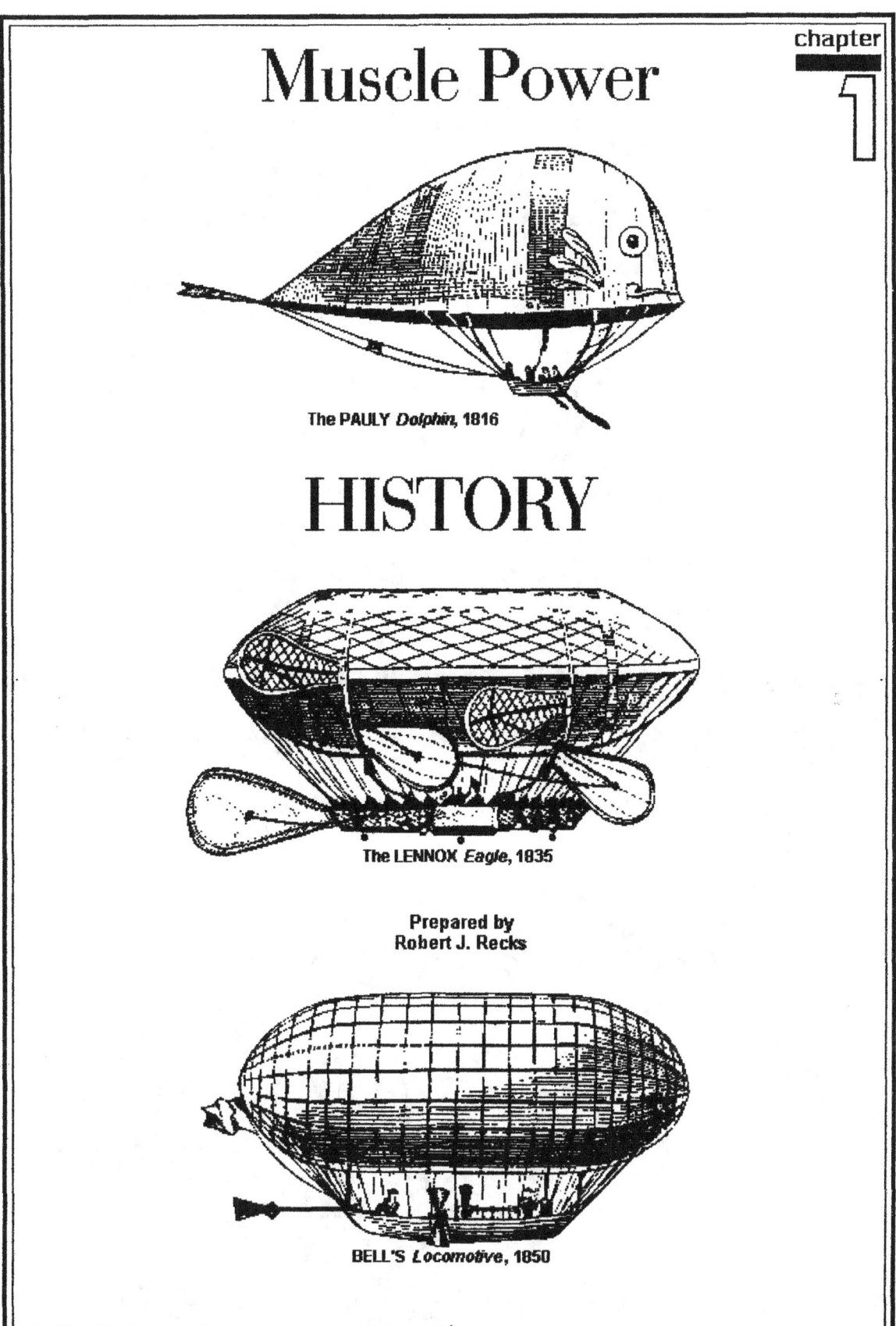

Muscle Power

HISTORY

chapter 1

The PAULY *Dolphin*, 1816

The LENNOX *Eagle*, 1835

Prepared by
Robert J. Recks

BELL'S *Locomotive*, 1850

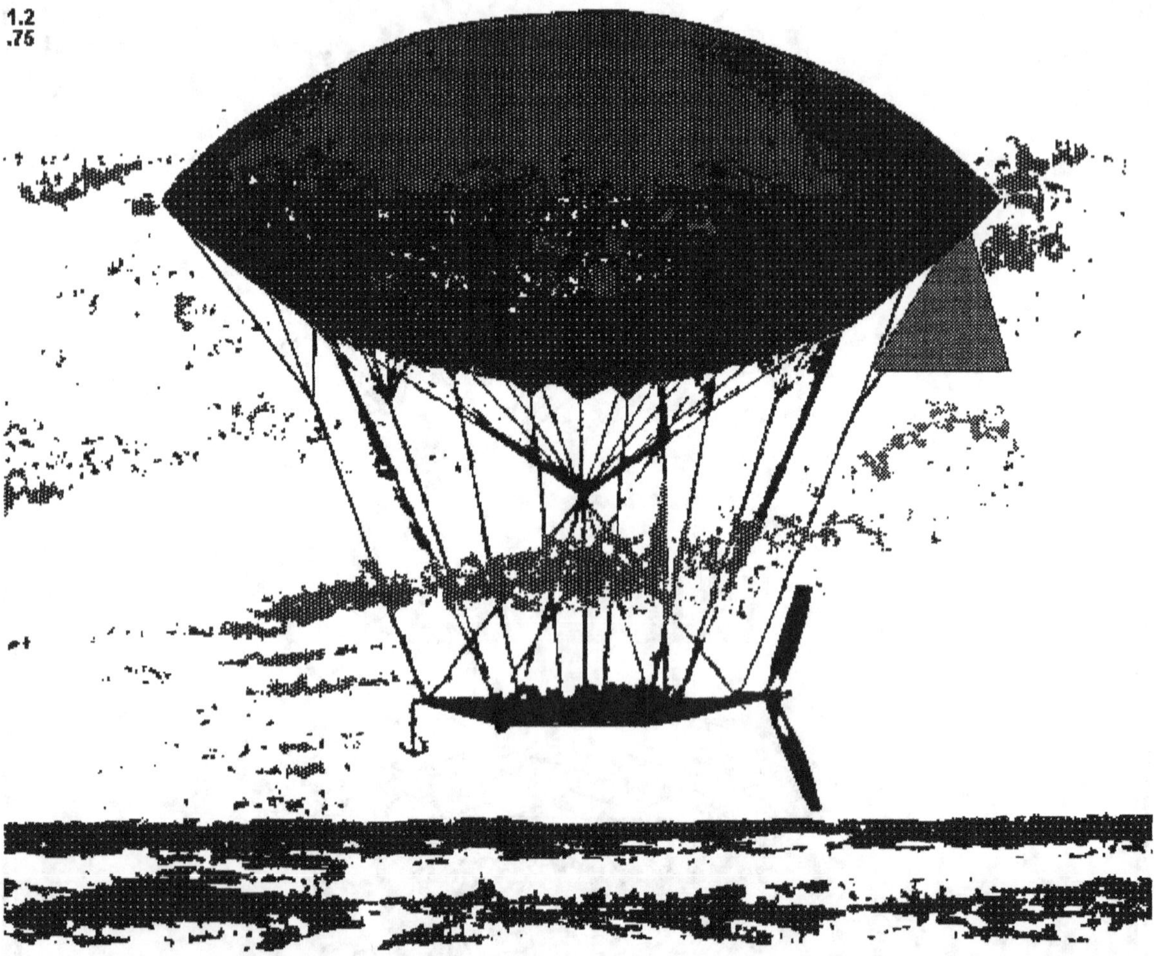

Dupuy de Lome's man-powered dirigible - 1872

HISTORY

Stanislas-Charles Henri-Laurent Dupuy De Lome (1816-1885)

He was a noted French naval architect who was a member of the government's Committee of Defense that took over the reins of government when Napoleon-III fell in September 1870. He was given a credit of 40.000 francs to build a giant steerable dirigible. Capitulation ended the project before the craft was completed, but two years later he felt ready to try again. This time he built a small dirigible, similar to Giffard's, measuring 108 feet in length and 47 feet in diameter.

With no lightweight engine available, he was forced to depend on manpower. In a large wicker car slung by a network of triangular ropes, he provided room for eight "galley slaves" to turn a long crankshaft linked to a 4-bladed propeller at the bow.

The Dupey de Lome dirigible made only one flight, from the fortress at Vincennes, on February 2, 1872, and proved to be somewhat successful by veering 12 degrees off course under power—defying a light wind. A safe landing was made at Mondecourt after achieveing an airspeed estimated at 5 miles an hour.

The aircraft had all the features of a modern non-rigid airship with control and stability, suspension system, ballonet and lower and an envelope of double-ply rubberized and doped fabric. Only an adequate powerplant was missing. Later in that decade two other balloonists, Albert and Gaston Tissandier (who had also built war balloons during the Siege of Paris), fitted an electric motor to a dirigible.

Complete Book of Airships, p34

The **HAND CRANK** *"Dirigicycle"*

Charles Francis Ritchell, a native of Portland, Maine, had unveiled an airship with a barrel-shaped gasbag powered by a hand crank as early as 1878. Ritchell was living in Corry, Pennsylvania, at the time he patented his airship. He was already well established as a mechanic and general inventor. His local fame was based on his invention of a boring machine used in the manufacture of brushes. He moved to Bridgeport, Connecticut, in 1878, where he was employed by the firm of Ives, Blakeslee, and Company, manufacturers of mechanical toys. There his inventive genius as given free rein. There were, for example, his walking dolls, a child, a sheep, and a cat "that has only to be pulled to walk off as naturally as can be, and is without clockwork, and needs no winding."

The list of inventions that poured forth from Ritchell's fruitful mind seemed endless. By the mid-1880s he claimed to have obtained 150 patents on such pleasant and harmless devices as windup toy motors, corsets, and envelopes with the gum on the body rather than on the flap. In spite of these many well-publicized inventions Ritchell remained best known for his airship experiments. In later years the inventor claimed to have begun thinking about the problems of flight in 1869. Nine years later (1878) he was awarded U.S. patent no. 201,200 for an "improvement in flying machines". His craft was first flown in public at Philadelphia's Concert Hall in May 1878.

The barrel-shaped gasbag was made of rubberized fabric 25 feet long and 13 feet in diameter, with a weight of 66 pounds. The envelope was so small that it could be flown only with hydrogen produced in the standard method of adding sulfuric acid to iron filings. Seven broadly worsted bands extended over the top of the gasbag and attached to a 23-toot frame of nickel plated mandrel drawn brass tubing 1.5 inches in diameter. The airship framework was suspended from this keel.

The operator's frame was constructed of the same brass tubing as the keel. It was 11 feet long and 2.5 feet wide and constructed in two sections. The pilot sat to the rear of the frame with his feet on two pedals and with a hand operated wheel crank between his knees. The hand crank turned a four bladed propeller set horizontally beneath the pilot. Twenty-four inches in diameter, it worked at a speed of 2,000 RPM. The four blades were cut from white hollow (oak wood), each with a surface area of 50 square inches.

A tractor propeller measuring 22 inches across was placed at the very front of the frame and was geared to the main horizontal propeller and hand crank drive system. It was controlled by the foot pedals. By pressing on the left treadle, the pilot threw the forward airscrew into gear, permitting it to operate at speeds of up to 2,800 RPM. When the pilot pressed his right foot forward the tractor propeller swung to the left; a push to the rear with his right heel pivoted it to the right.

The finished machine weighed only 114 pounds and was normally flown 6 pounds heavy. No valves or ballast were employed. To rise, the pilot turned the hand crank forward to operate the lifting airscrew. To descend, he spun it in the other direction. The tractor propeller was employed for forward motion and for turning. The

The Eagle Aloft p493-900

The *DIRIGICYCLE*

1.2
0.6

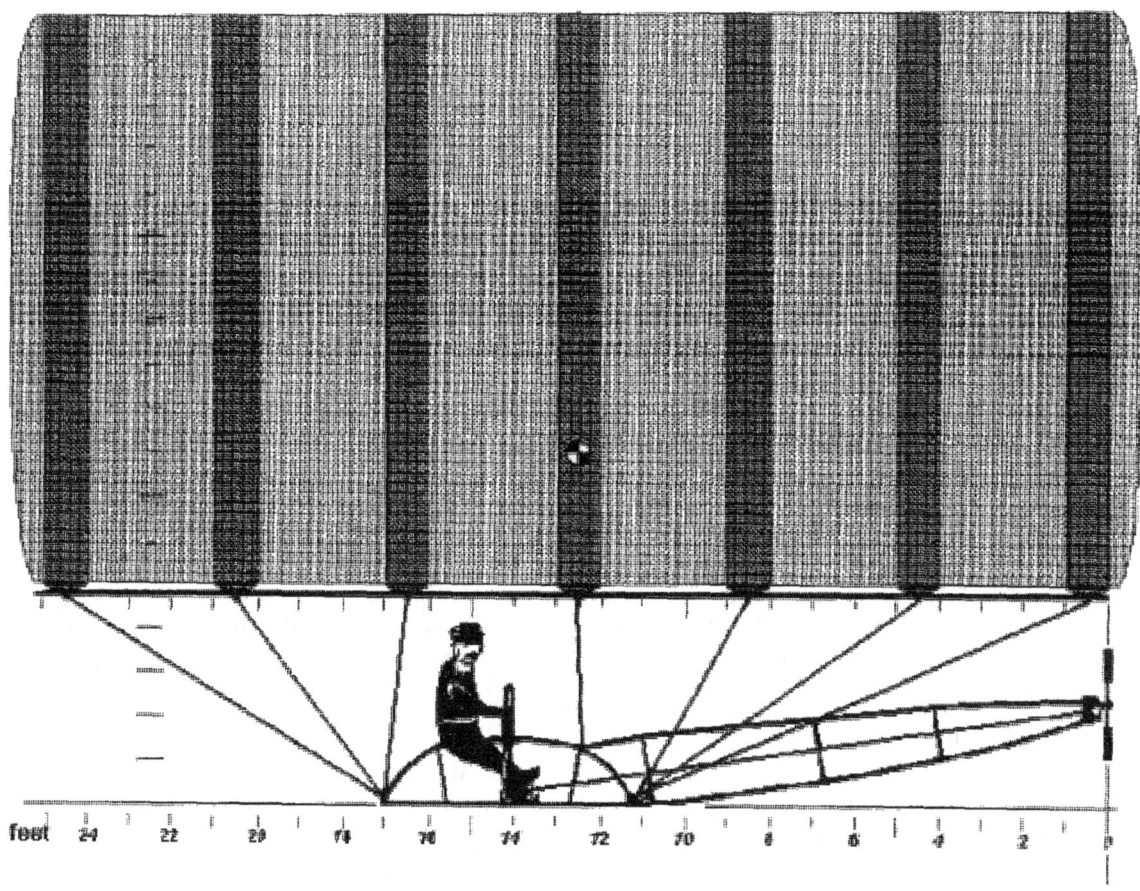

Charles F. Ritchell, *Inventor*

The *DIRIGICYCLE*
Specifications:

Envelope: Length = 25' Diam. = 13'
 Volume = 3369 cuft. Hydrogen
 Material = Rubberized Cotton
 Weight = 66 lbs; Lift = 232 lbs.*
Gondola: Length = 17'; Width = 2.5'
 Structure = 1.5" Brass Tubing
 Weight = 48 pounds
Horizontal Power = Hand Crank
 22" Diam., 4-Blade wood Prop;
 50 sqin. area, 2800 max RPM.
Vertical Power = Hand Crank
 24" Diam. wood, 4-Blade Prop;
 50 sqin. area, 2000 max RPM.

* Asuming 69#/1000 cuft.

Ref. Patent # 202200 (1878)

Early Muscle Power Successes
Charles Ritchell, cont.

inventor, who was too heavy to fly the craft himself, hired a lightweight young man for the early indoor exhibitions. Not long after the beginning of his successful run in Philadelphia, he was deluged with young women volunteers answering his advertisement for a female pilot. He finally selected Miss Mabel Harrington, who, suitably clad "in her new suit resembling a silk jacket and knee breeches," made a number of circles of the hall to the applause of an admiring crowd.

The first outdoor trial of the Ritchell airship took place in Hartford, Connecticut, on June 12, 1878. The pilot, 96-pound Mark Quinlan, rose to 250 feet, sailed by the spire of the Colt Memorial Church and over the Connecticut River. Quinlan had ascended in a dead calm, but a strong wind sprang up while he was in the air, presaging the onset of a storm. The pilot was able to crank his way back to the ball park where the flight had begun, landing a few feet from his take-off spot before the rain began.

The following day, June 13, Quinlan once again demonstrated that he had "the confidence and nerve enough to go up in a gale." A great deal of time was lost in properly balancing the machine that day. Loaded with nine pounds of stone ballast, the craft rested lightly on the ground, but could be lifted with a slight pressure of one finger. A local reporter described the take-off: "Then the word was given to 'Go.' Quinlan began turning the wheel, the horizontal fan revolved with a noise like a buzz-saw, and the machine darted vertically to a height of about two hundred feet. There a steady current of wind setting toward the southwest was encountered, and the machine was swept away by it.". Operating the forward propeller at maximum speed, the aeronaut could not make headway against the wind. He dropped to 100 feet, then rose to 300' apparently searching for lighter winds, all the while being blown out of sight toward New Haven. After nearly an hour in the air, Quinlan gave up his attempts to return to Hartford and landed at Newington, 5 miles from the take-off point.

Ritchell began exhibiting his machine at Boston's Tremont Temple on June 24, 1878. The demonstration, arranged by William McMahon, who played a major role in introducing Edison's phonograph to the public, was a complete success. As the Boston Herald noted, "It is hardly an exaggeration to say that a more novel and interesting exhibition has never been given in this city. In addition to the indoor flights, Quinlan made an exciting ascension from Boston Common. Once in the air, the propeller gears jammed, allowing the balloon to rise dangerously high. Without a valve, the envelope swelled, breaking several of the bands from which the frame was suspended. Quinlan could not slit his envelope, for there was no netting in which the fabric could gather to form a parachute. He had little choice but to tie one hand and ankle to the frame, then drop beneath the craft to make repairs with a jackknife as his only tool. He finally descended at Farnumsville, 44 miles from the Common, after a flight of one hour and twenty minutes.

Business was looking up for Ritchell. He was receiving orders for airships from other prospective exhibitors, including one Cuban syndicate. In cooperation with the D.P. Ells Company of Chicago, Ritchell marketed toy airships known as Patent Flying Wizards. Selling for fifty cents each, the Wizards were guaranteed to fly and "furnish splendid outdoor exercise and amusement to old and young, boys and girls.

The Eagle Aloft p493-900

Early Muscle Power Successes
Charles Ritchell, cont.

Mitchell, Quinlan, and the airship also found employment with W. C. Coup's Greatest Show on Earth in 1878. Ritchell continued to operate his original machine at Brighton Beach, near Coney Islands in 1879. A larger machine known as *Peerless*, apparently constructed to carry two persons, broke loose and blew out to sea. The inventor was also said to be building a very small airship for W. C. Coup. Weighing only six pounds, it would be operated by a dwarf. By 1880, as the New York Times noted, "even flying machines lose their novelty. Ritchell was now regaling editors with the prospect of an aerial voyage to the North Pole. He was also giving some thought to the possibility of a rotary wing, heavier-than-air machine and to the construction of an 83-toot-long version of his small airship. With such a craft, he predicted, flights of up to twenty-four hours would be possible. He was entirely pessimistic, however, about the ability to fly against the wind. "After all my experiments of building five complete flying machines and working them, I am convinced that if an engine could be made of 1,000 horse power weighing but one pound and no larger than your hat, it could not propel a balloon of 3,000 foot capacity against a wind current of ten miles an hour five miles an hour.

Ritchell was never to alter his attitude. As late as 1905, long after the appearance of very large Zeppelin airships, he would still remark that the possibility of navigating an airship against the wind remained "just as ridiculous as a perpetual motion machine, and the latter will be invented just as soon as the former." Still an adherent of hand propulsion, he did believe that the crew of such a craft would be able to search out favorable winds blowing toward their destination. Ritchell's rather primitive notions of aeronautical propulsion guaranteed that he would make only minor contributions to lighter-than-air technology. He was, nonetheless, a superb showman who inspired other, more experienced, aeronauts to attempt their own airship building programs. The moral of the story seems to be:

Invention is Open-minded Innovation

The Eagle Aloft p493-900

MUSCLE POWERED BLIMPS

The <u>PEDAL POWERED</u> *Skycycle*

Carl Myers was the best-known experimenter who followed Ritchell's lead in constructing airships. Though there is no evidence that Charles and Carl ever met, it is highly likely that they did cross paths in the carnival circuit. But Carl chose to depart from "muscle-power" of his arms, to the "pedal-power" of a bicycle, and called it a "Skycycle." It was an aptly named single seat aircraft, to the delight of thousands of aeronautical enthusiasts between 1879 and 1900.

Admittedly, it was originally loosely based on Ritchell's designs, but Myers quickly developed a few sensible touches. His craft were not only pedal powered, but the early Skycycles featured both a rudder at the rear and forward steering vanes. The rudder was abandoned in later versions. A shift of the pilot's weight was sufficient to climb or descend. (68)

Carlotta, his wife who spent a great deal of time on the seat of a Skycycle, described the appearance and operation of one of the early "gas kites":

> Looked at on its side it appears like a canoe bottom up. Viewed in front its section seems like that of a half sphere or dome. Its under side is flat, like a kite, and its interior space is filled with gas enough to nearly lift the entire machine and aeronaut A balloon netting surrounds it, and its cords support a concentrating ring exactly as with an ordinary balloon. Instead of the customary basket, there hangs from this ring a velocipede seat. In front of the operator, where the "steering bar" of a velocipede is, there are cranks for the hands instead. At the feet are ordinary velocipede cranks. All are geared so that moving one crank moves all, and together they revolve a screw shaft which projects to some distance in front like a bow sprit, supported by the netting stays. At the outer end of the shaft is a huge screw of cloth supported by two yards, like a ship's sail. This lies flat and motionless like the outspread wings of a soanng bird until revolved, when it instantly twists itself into a screw.(69)

The exact number of Skycycles constructed is uncertain, but the couple flew exhibition dates with the machinery in thirteen states. The craft were so simple to operate that novice pilots were occasionally allowed to take a hop. Such was the case on August 3, 1895, when a New York World reporter pedaled his way to the Brooklyn Navy Yard across the East River and Manhattan to a landing in Yonkers.(70)

Carl Myers finally patented the Skycycle in 1897. Together he and Carlotta continued to operate the small airships into the early years of the new century. In 1900, for example, they made 120 Skycycle ascents during a single engagement at the St. Louis Coliseum. Myers's Electric Aerial Torpedo also developed by 1900, was a smaller, unmanned version of the Skycycle intended to deliver a load of explosives against an enemy.(71)

The Eagle Aloft p497-498

MUSCLE POWERED BLIMPS

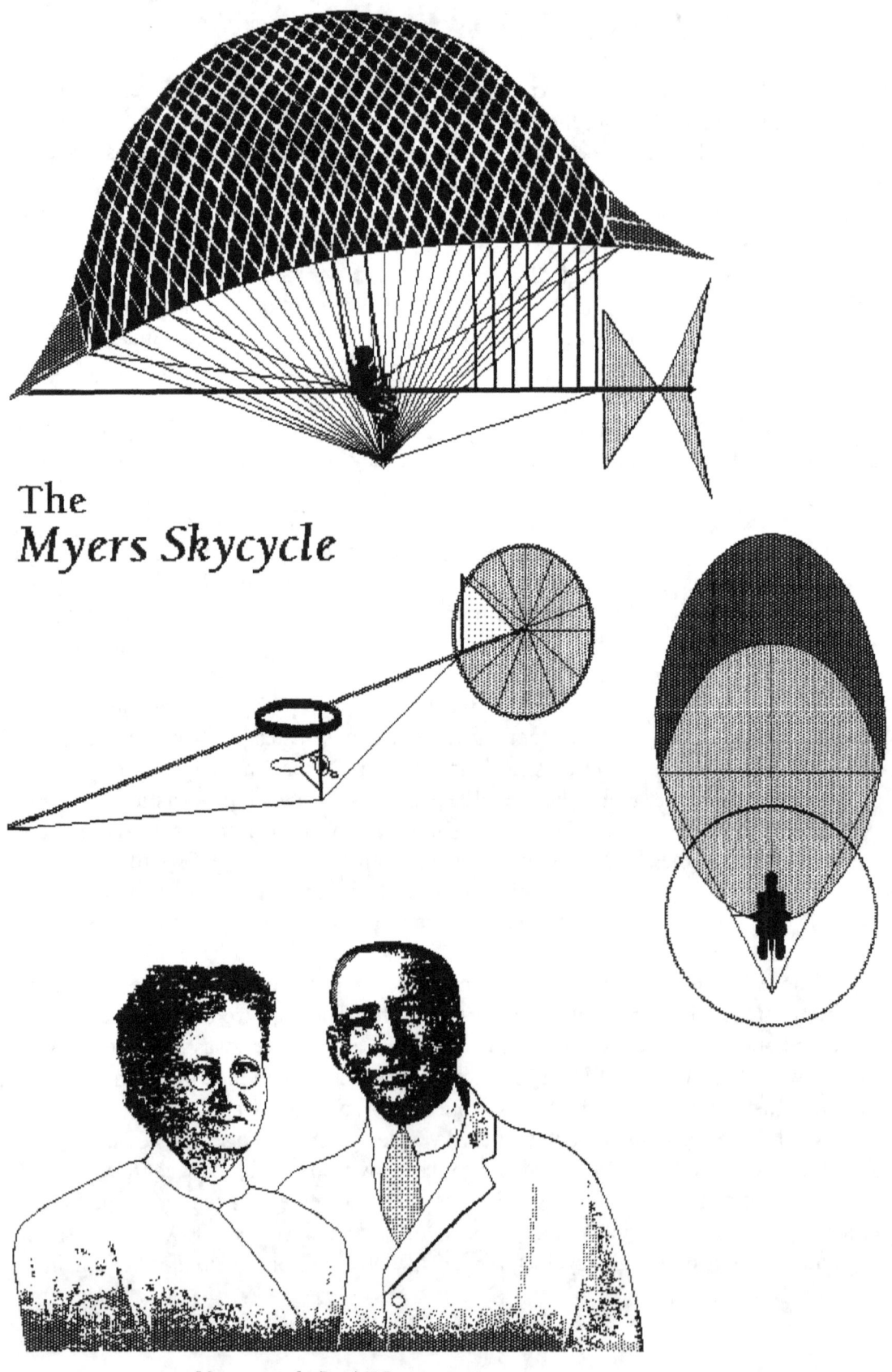

The *Myers Skycycle*

Mary and Carl Myers

MUSCLE POWERED BLIMPS

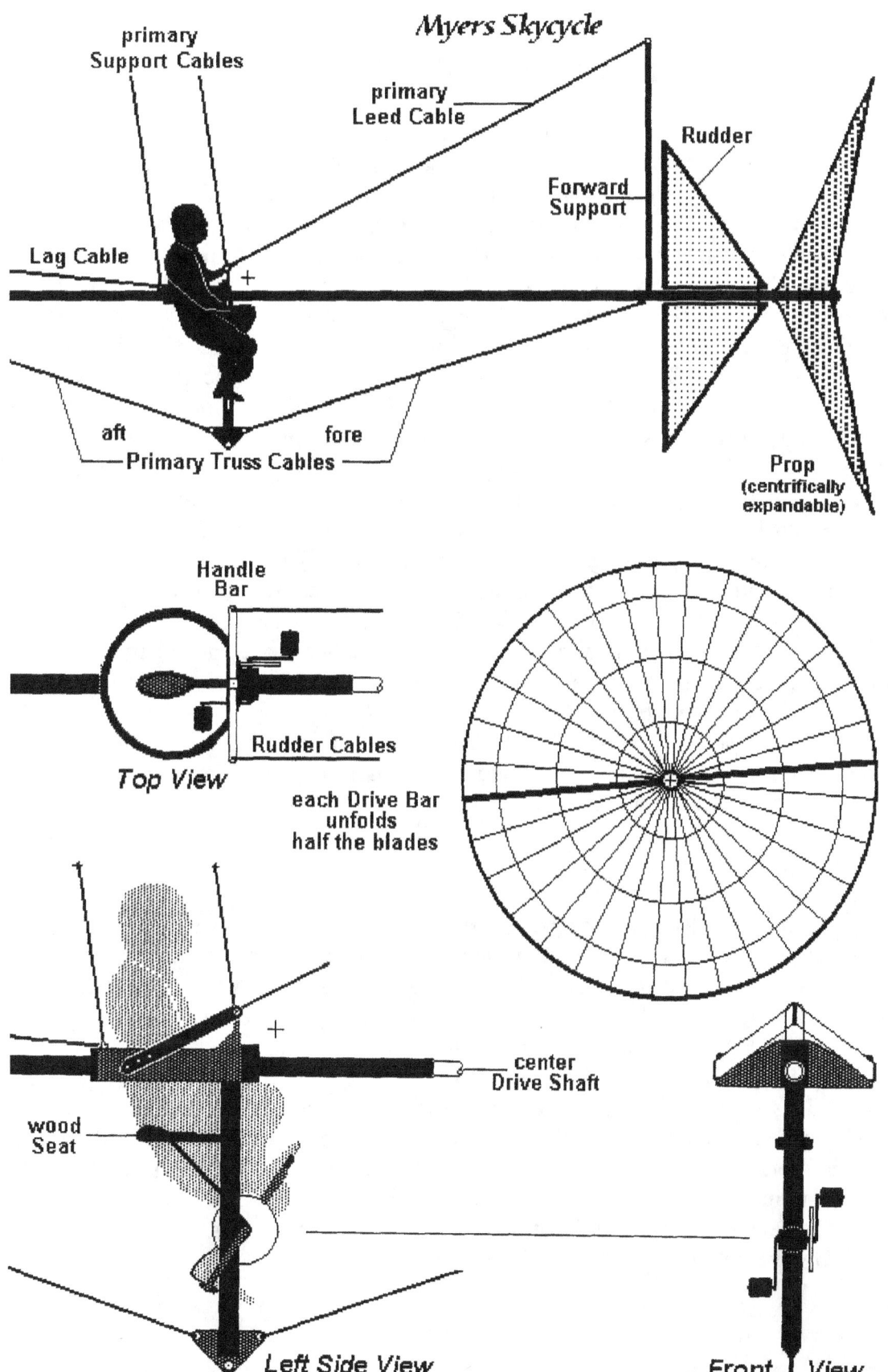

The **PEDAL POWERED** *"Campbell Airship"*

While Ritchell and Myers were the best-known aeronauts operating human-powered airships, a few other balloonists followed suit. Arthur Bamard. of Nashville, Tennessee, for example, made a number of pedal-powered flights in 1897.

The Campbell Airship, one of the more controversial projects, was the product of a number of the major aeronauts of the period, including both the Myers and Allen families. The airship was conceived by Peter Carmount Campbell. A native of Rhinebeck, New York, born in 1832, Campbell had for many years operated a jewelry store in Brooklyn. A longtime aeronautical enthusiast, he had discussed his schemes with Horace Greeley, S. F. B. Morse, and others for many years before beginning construction of his airship in 1888. The craft was built around an oblong envelope 60 feet in length and 42 feet across. As with the Ritchell machine, a long metal rod beneath the gasbag served as a keel. The keel was directly tied to the bag at the center. A web of cords extended from the bar to the ends of the bag. A boatlike car was slung from the keel. Large enough to house a pilot and three passengers, the car sported two large, birdlike wings on each side. The wings were not designed to be flapped. but could be raised or lowered to control the direction of motion. A forward rudder was also employed.

A large, multi bladed propeller was located on the underside of the car. Campbell originally hoped to power this fan, and the three smaller pusher propellers at the rear, with batteries. In order to save weight, however, the electrical power system was rejected in favor of foot pedals. (72)

In August 1888 Campbell contracted with Carl Myers for the construction of three small practical working models and a single large envelope and net for a full-scale airship. Campbell and the investors in his airship enterprise worked the rest of that summer and fall to prepare their craft for flight. By early December 1888 they had erected a gas generator in a vacant lot next to the Sea Beach Palace at Coney Island. New York, and were ready for a trial. (73)

James K. Allen was engaged to operate the craft. Allen, worried that the freezing temperatures might crack the fabric, waited until December 8th, a warm day, to attempt an ascension. After several hours lost in generating hydrogen, Allen climbed aboard at 4:00 P-M, tested the pedals and rudder rope, and ordered his release. Ascending to 100 feet, the aeronaut heard Campbell calling him to return to earth for a photograph. Allen cranked himself back down without difficulty.

Reascending to 500 feet, Allen paused to gauge the strength and direction of the wind. He then flew toward Brooklyn, reversed course back to Coney Island. and spent the next half hour zig zagging across the skyline. "I never had a pleasanter sail in my life," Allen informed reporters. "This mechanism is nearly perfect, and the ship minds her rudder as quickly as does a sloop at sea. The labor of propulsion is not at all difficult. I wish I owned the ship. 1 could make a fortune out of her. (74)

The Eagle Aloft p499-500

Early Muscle Power Successes
Peter Campbell, cont.

The Campbell Airship was taken back into storage until the following spring. Campbell had engaged the services of Canadian balloonist, E. D. Hogan, to operate the craft in 1889. Hogan had apparently met Campbell while making regular parachute drops from a hot air balloon over Rockaway Beach in the summer of 1888. Campbell and Hogan made their first attempt to launch the airship on June 19, 1889. After five hours of effort, they were forced to admit defeat to a jeering crowd of 1,000 spectators. The failure to ascend was blamed on a leaky gas generator.

Charles Ritchell, who had followed Campbell's progress with interest, was incensed. Campbell, he charged, had directly copied his own airship plans. "I abandoned the machine," Ritchell complained to the press, "because it was only a toy, and couldn't be made of any practical use, but I want whatever glory there is in the invention and don't propose to let these people go on deceiving the public with a machine built after my ideas and which is just as useless and even more helpless than mine was. (79)

Campbell ignored Ritchell's comments and brought the airship out for another trial on July 16, 1889 On this occasion, Campbell and Hogan abandoned the old generator in favor of city gas The ascension began well, with Hogan moving toward the southeast. When the ship had traveled roughly a mile from its takeoff point at the Nassau Gas Company works, the large horizontal propeller beneath the car was seen to fall into the water.

Some observers thought they could see Hogan climbing the network as the airship was blown out of sight Late that afternoon, the crew of a New York pilot boat reported seeing the balloon dragging in the water 74 miles off Long Island. The boat was unable to overtake the balloon. No trace of the aeronaut was ever found. (76) The moral of the story seem to be:

The Wind is NOT your Friend

The Eagle Aloft p499-500

The **HAND POWERED** *"Canada Airship"*

The desire to navigate the air had infected a few of Hogan's countrymen North of the border. One of the best-publicized airship projects of the period was the work of two citizens of Montreal, Richard Cowan and Charles A. Page.

Page, "a very ingenious and skillful mechanic," began to consider the possibility of the navigable balloon during the Franco Prussian War, when free balloons were employed by besieged Parisians to communicate with the outside world. (77) Page's invention lay fallow until 1878, when his work came to the attention of Richard Cowan, a wealthy retired Montreal merchant. It would seem that Charles Ritchell's limited success with his pedal-powered airship that summer was a factor of some importance in drawing Cowan's attention to the subject.

Fortunately for Page and Cowan, neither of whom had any aeronautical experience, Charles H. Grimley, a veteran of forty-one balloon ascents, was on hand to take charge of practical details. He began a long Canadian balloon tour in March 1878. Over the next eighteen months he made four ascents from Ottawa, the last on the occasion of the annual picnic of the St. George Society on August 20, 1878. He narrowly escaped falling into the St. Marys River after an hour in the air, but the crowd, sponsors, and press loved it.(79)

Cowan and Page displayed the partially completed car of their proposed airship at the Shamrock Lacrosse Gardens on the occasion of Grimley's flight. The relationship between the three men apparently began at this time. By the following spring the car was complete. and so was a brand new 70,000-cubic foot aerostat which Grimley had commissioned for Cowan and Page. The first flight of the new craft was scheduled for another Irish Protestant Benevolent benefit on June 21, 1879.

The Cowan-Page car was the subject of some attention in the Montreal newspapers. The passengers would travel in a cage of iron tubing 7 feet long, 7 feet high, and 4 1/2 feet wide with a plank floor. Twin side wheel propellers driven by hand cranks were attached to the side of the passenger car. The car was enclosed in a larger iron frame, 30 feet long, designed to be slung beneath the usual cigar-shaped gasbag. .R single-piece, cruciform rudder-elevator was attached to the rear of this frame. Provision was also made for a valve on the balloon, although it was hoped that, through judicious use of the paddles and control unit, it would not be necessary to resort to valving gas or dumping ballast.

For the preliminary experiments, Grimley's new balloon *Canada* would be substituted for the ideal cigar-shaped envelope. Four times the size of the old City of Ottawa, the new craft stood 80 feet tall and measured 50 feet in diameter.(80)

The premier ascent of the new balloon and car on June 21 was a disappointment to Cowan and Page. Delays in the inflation forced Grimley to cut the process short. He had originally intended to fly with Cowan, Page, and two reporters, a Mr. Creelman and Hiram Moulton. Now he asked Cowan and Page to give up their seats in favor of one of the journalists in order to foster publicity for the program. The craft was still too heavy

The Eagle Aloft, p501-503

Early Muscle Power Successes
Canada Airship, cont.

however. Moulton was next to go, followed by the fore and aft sections of the car, the propelling wheels and mechanism. Grimley and Creelman would make a normal balloon voyage using valve and ballast.

It was a rough journey during which the Canada was blown over the Richelieu and St. Lawrence rivers toward a vast forested area beyond. On landing, area before their grapnel took hold. Grimley was knocked uconscious but neither man was seriously injured.(82)

Grimley moved on to Ottawa to entertain at the Dominion Exposition that July, while Cowan and Page were reassembling the car for another trial. During this waiting period, Charles Ritchell's airship became a subject of discussion in Canadian newspapers. One disappointed correspondent, who had attended Grimley's ascent on June 21, noted that while Ritchell's machine was smaller, it had proven far more trustworthy.

Anxious to redeem themselves, Cowan, Page, and Grimley scheduled another test flight for July 31, 1879, from the Shamrock Lacrosse grounds in Montreal. On this occasion all three men flew with three additional passengers Mr. Moulton, ton of the Montreal Witness, Mr. Browning of the Montreal Herald, and Mr. Harper of the Montreal Star.(83)

All navigating equipment was in operating order this time and Moulton was ordered to take the helm soon after take off. After enjoying the scenery from 1,500 feet, Grimley ordered his crew to man the sidewheel propellers. Grimley and Harper operated the windlass on one side while Page and Browning manned the other propeller.

Everything seemed to be going well until Page noted that the revolving propellers were cutting the suspending ropes connecting the car to the balloon. After some minor repairs they set the propellers in motion a second time, only to start up a bumping and jolting in the car. The Camp reverted to the status of a free balloon once again while Grimley climbed out of the car to effect still more repairs. In the third trial, the propellers functioned as planned. While the equipment proved inadequate to control the ascent or descent of the craft by itself, it did help to conserve gas and ballast.

The mechanism created one anxious moment when the valve and rip line became tangled around one of the cranks. If the ripping line were pulled, the entire top of the balloon would open, allowing all the gas to spill out. Afraid to attempt disentangling the lines, Grimley began searching for a landing place. After two and a half hours in the air, they came safely to earth near the village of St. Aimee.

Cowan and Page were more than satisfied with the outcome of this final voyage. While their equipment had certainly created problems it had proved effective in stabilizing altitude. Others were less certain. Correspondents writing to the *Montreal Star* continued to call attention to the fact that Ritchell's small machine had operated with greater success and fewer problems. It did show that:

Bigger is Not Necessarily Better

The Eagle Aloft, p501-503

The CENTENNIAL

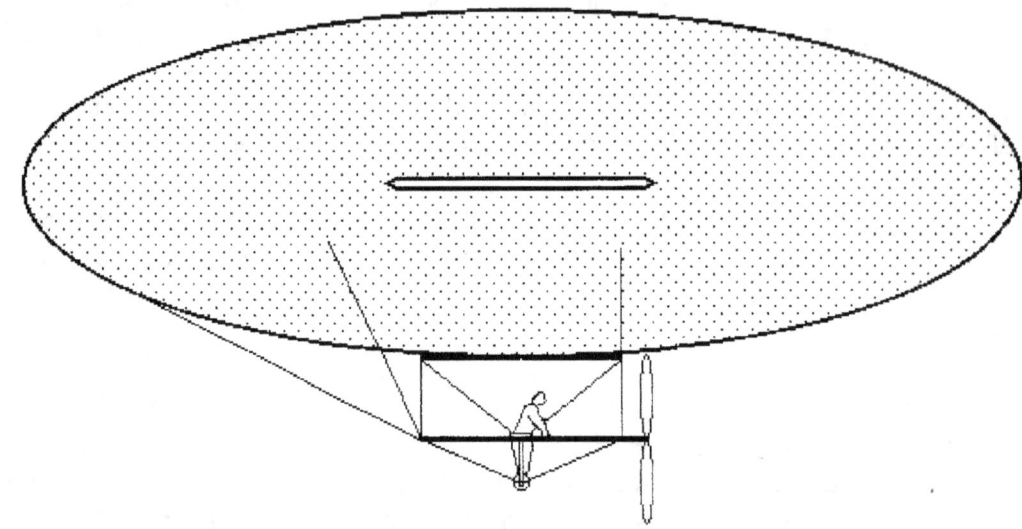

Arthur Barnard and the Centennial

The bicycle of the sky was patented by produced many "skycycles" in the 1890's. Professor Arthur W. Barnard, Director of Physical Training for the YMCA of Nashville, assembled an egg-shaped airship measuring 18 feet in diameter and 46 feet in length. It was filled with hydrogen and used a propeller of 8 feet in diameter.

Two steering surfaces, called aeroplanes by the aeronaut, were each 10 feet square and located at the equator on either side. The hull (envelope) was larger than the one-man "skycycles" built by Charles F. Ritchel or Carl E. Myers.

Barnard pedaled his *Centennial Airship* in Nashville at the Tennessee Centennial Exhibition of 1897. He went a distance of some 20 miles with the help of a strong wind aloft. But on the return the spar broke off one of the propellers and he landed twelve miles short of returning to the Exhibition grounds.

The Nashville Daily American

The Wing Powered Flying Machine

While most aeronauts are discouraged or preoccupied with the problems of flight, the well-known Russian aeronaut, Dr. Danilewsky, was lead to make this statement: If our body was less heavy, it would be much easier to steer it in the air. The largest amount of the effort of a flying man being employed to remain aloft. (He said "Let me be lifted and held aloft and I will know how to steer.")

This was the reoccurring problem of the steerable airship, but bringing a new solution, to apply to aerial navigation a combination analogies to that of "The blind and Paralytic of La Fontaine". The balloon would carry, and the wing would steer. See Picture B - The flying machine of Dr. D. The aeronaut before setting off, attaches his wings to the balloon.

This strange aircraft is stretched in the form of a cigar. to which is linked a pair of wings near to 10 meters in span. It is with this little balloon, inflated with pure hydrogen, that Dr. D rose. our strange illustrations represent the Russian aeronaut at the time he carries out, at Charkoff! One of his experiments, before a detachment of military balloonists. The air is calm. The balloon advances, turns, descends, and rises, while underneath him the enormous wings flap.

It looks like an immense white bird whose black head is the man. From below, the spectators see this head move. It is the aeronaut who flaps arms and legs to activate his **wings.** If one of these ropes which hold the small saddle where he is conveniently installed - wow, one must not have vertigo - if one of these strong breeches which hold him suspended in the open should break and, like Lilienthal, the aerial voyager is assured of a fatal fall.

It is different than the airship that Mr. Carl Myers, inventor of a curious attempt made in America, uses to rise into the air. But he doesn't attach Dr. D's wings to his balloon. It is . a Velocipede. Do not smile. The aerostatic velocipede has proven itself.

Seated on his machine, at 400 meters altitude, the aerial cyclist has covered quite a long distance, making his balloon rise, descend, turn and incline itself. To be noted: the balloon weighs a little more when fully loaded, than the volume of air it displaces. It is the action of a directive propeller, fixed above the velocipede, which maintains it floating at the same time allowing it to turn as desired. If the prop ceases to function, Mr. Myers lets go of the pedal, and the balloon starts to descend immediately and if the pedal is not resumed in time the pedaling aeronaut runs the risk of a terrible fall, unless he uses his parachute.

Man fighting the wind

Despite the ingenious attempts, the capital discovery remains to be made. Neither the rowing balloon of Dr. D (both wings he commands are in fact but the oars of an aerial

MUSCLE POWERED BLIMPS

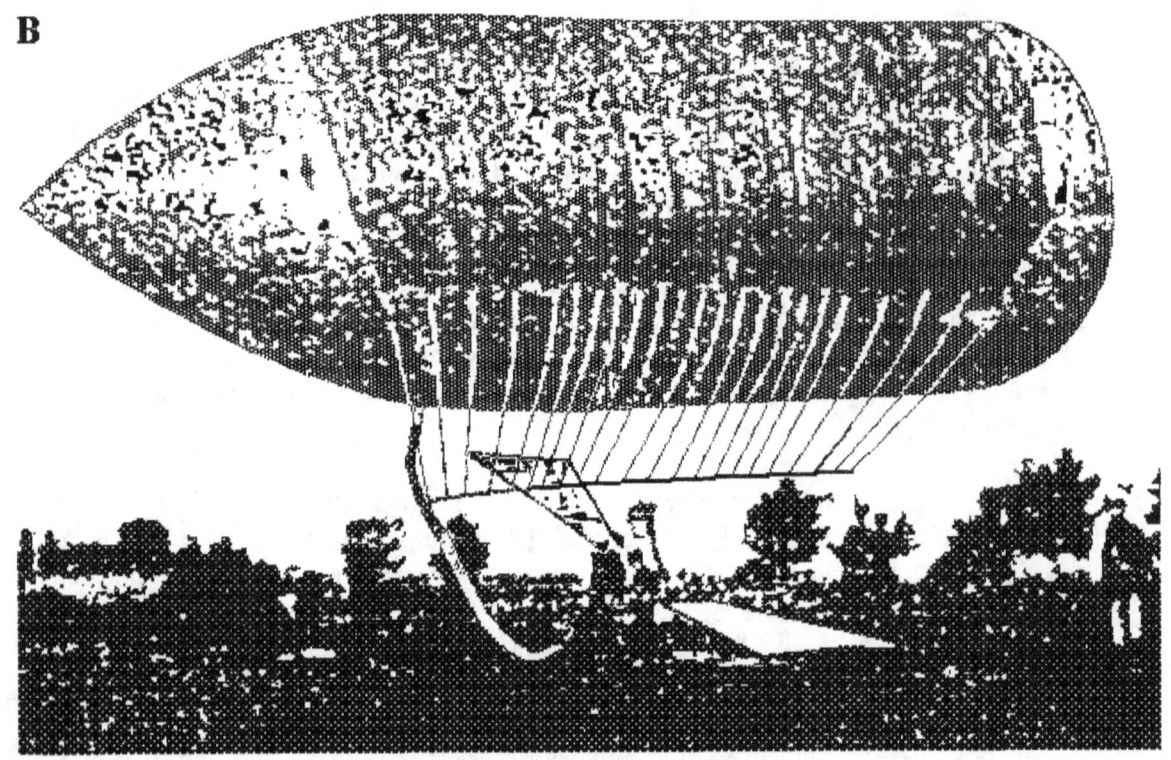

Doctor Danilewsky

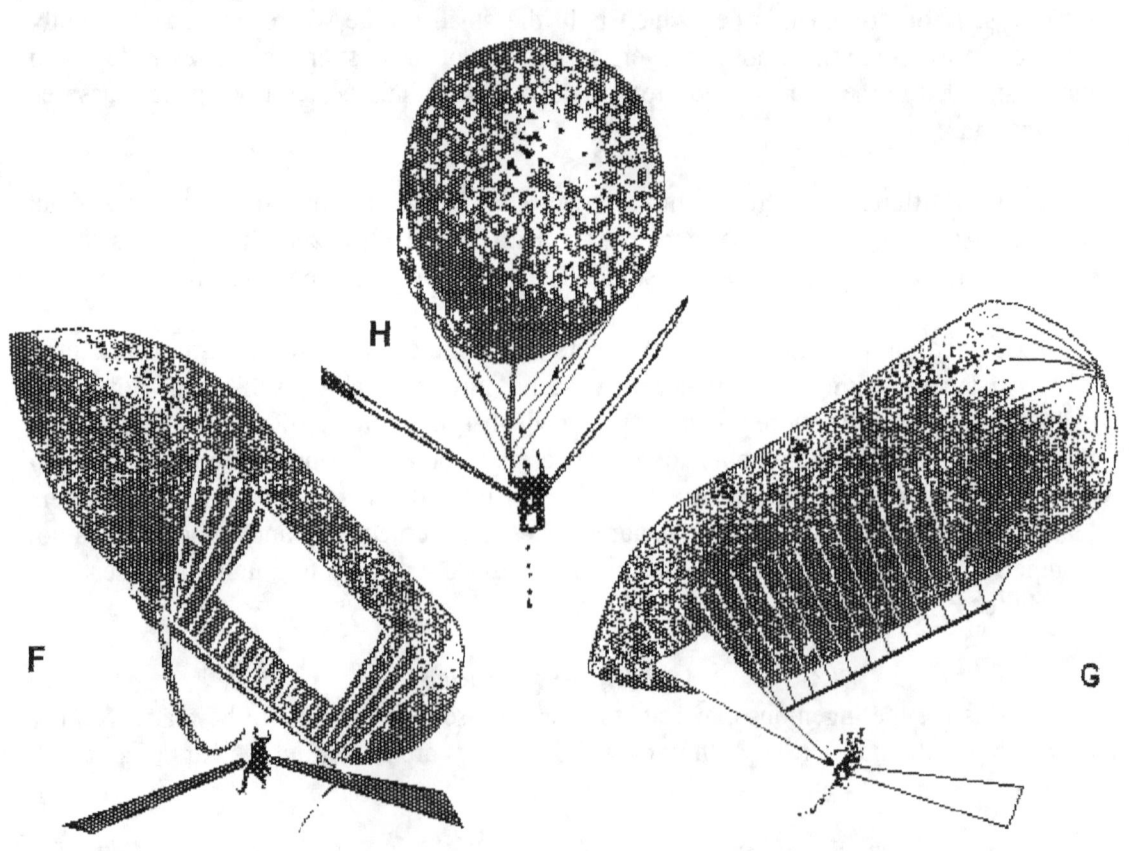

Early Muscle Power Successes
Dr. Danilewski cont.

boat), nor Mr. Carl Myers' velocipede can claim to become common machines. They always have an enemy which they must always account for and against whose anger they cannot easily battle.

This enemy is the wind, the terrible rebel who flouts the domination of man. By necessity the Russian scientist and the American inventor must ascend in calm weather. Do you see them venturing out without consideration, the one with his immense wings with complicated mechanics and offering the lightest storm a surface of resistance which would make the shock more violent, the other with a velocipede soon overcome by the wind? For an aerostat steered by some propulsion system to progress as intended, i.e. counter the wind, and not be driven along with it, the speed reached must be greater than that of the aerial wind blowing in the covered areas.

The last machine to fly.

The dirigible aerostat of Danilewsky, maneuvering in the air. To steer, the aeronaut activates his wings orienting them differently to climb, descend or glide, turn left or right. Well, is there an engine capable of giving a balloon the speed of the air currents? Even without talking in storms, which blow between 30 and 40 meters per second, i.e. 100 to 150 km/h1 almost twice the speed of express trains, we find that the average wind speed is 10 meters per second, i.e. 35 km/h. There is the question. The moment the motor capable of successfully combating these winds is found, aerial navigation will become a game within reach of everyone.

However all that has been attained is a speed of 6 meters per second. This is the speed of the wind on calm days. This was found in 1885 by the leading officers of the Meudon aerostatic park, Masseurs. Krebs and Renard, during assembly of their dirigible La France. If the wind had got up, the aerostat would cease to respond correctly and their test would fail, like that attempted last year by the German engineer Schwartz, with the dirigible he invention.

Fourteen metres long, entirely constructed of aluminum, this aerostat of the same shape as that of Dr.D.. i.e. resembling a giant cigar. sported a motor driving three propellers it took of at Templehof. near to Berlin, in a wind of 7 to 8 meters/second. It was supposed to be able, according to its inventor, to counter winds of 10 meters per second. On climbing to 180 meters altitude, it did indeed remain there for a few moments. But it was seen to soon loose shape, turn on itself, crash to the ground whereupon it was damaged. It had cost 200.000 marks, i.e. 250.000 Fr.

What fate is reserved for these latest flying machine? Like the ancients to symbolize the impossibility for man to go against nature, had imagined the Icarus legend. Carried by wings attached to his body by wax. Icarus rose too high; too Close to the sun whose rays had cemented the wax and the aeronaut plunged into the sea, which is now called the Icarian sea. However they did not dream of the thousands of discoveries of modern science which they would have found equally unbelievable. And we also would

Lectures Pour Tous, Paris 1899, 2' Jaargeng; pages 610-13

Early Muscle Power Successes
Dr. Danilewski cont.

most famous comedies, The Birds, the poet Aristophanes had imagined carrying the set in the city which the winged gentry forms between sky and Earth. Perhaps the day will come when man can become a citizen of the city of the birds. Let us simply say that this day appears very far away. In addition we do not see the practical advantages this discovery would have, whereas we do see the enormous difficulties. What has until now opposed the success of all aviation attempts, is that to be in ratio to the weight of man, the wings must deploy indefinitely, thus giving more and more susceptibility to the wind. And that is why these enormous, fragile wings which would lift to the cloud region, man created to live on the land. where all seems to be the wings of the fire dragon.

ILLUSTRATIONS

Picture B - Side view with the aircraft still on the Ground.
The flying machine of Dr.D. The aeronaut before setting off, attaches his wings to the balloon.

Picture D - The crew displaying various airfoils.
Many winged airfoils were tested. Some were 10 meters long and had an area of 26 square meters (282 sq.ft.).

Picture E - The aeronaut preparing for flight, surrounded by dignitaries.
The launch of the Danilewski dirigible at Charkoff. It required a maximum effort by the aeronaut to get in the air. The dirigible is directed solely by the use of the rowing wings.

Picture F - Climbing View of its left side. No original text was included.
Note the 16 suspension points on each side, and a banner on the left side only. Message unreadable.

Picture G - Climbing View of its right side.
Dr.D's dirigible maneuvering freely in the air. The pilot uses the wings to steer the dirigible by changing their orientation In this way he can climb, float, turn right or left, and descend.

Picture H - Front View from below.
The spectators see the dirigible climb and descend with the enormous wings are flapping around it.

Lectures Pour Tous, Paris 1899, 2nd annual volume, pages 610-12

DES AILES DE 10 METRES D'ENVERGUIE

LA DEPART DE BALLON DE DANILEWSKI A CHARKOFF

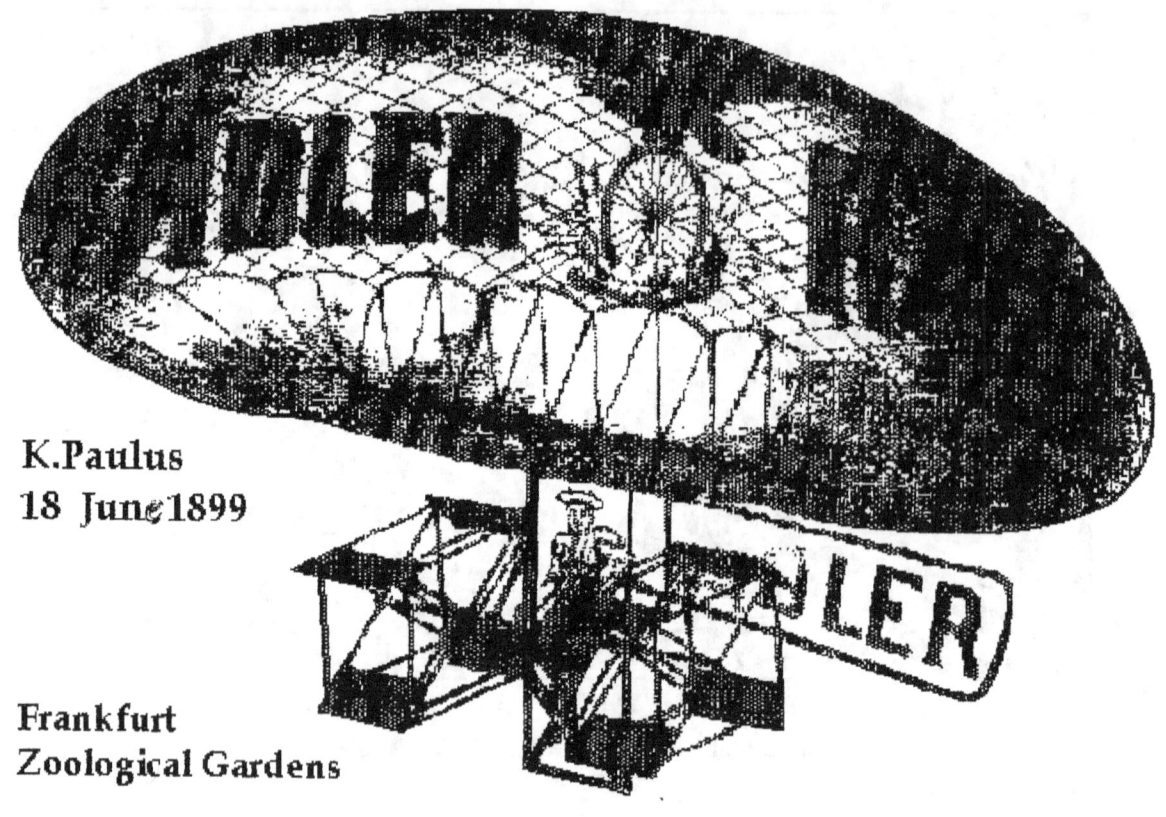

K.Paulus
18 June 1899

Frankfurt
Zoological Gardens

Katchen Paulus (1868-1935)
and the *ADLER-RAD*

Frau Paulus is still considered near the top of the list of pioneers in German aviation. She made her first ascent in 1893, witnessed the death of her fiancee (H. Lattemann) in 1894, and continued the business as the first professional woman aeronaut in Germany. She rode horses & Bicycles aloft, flew airplanes by 1910, and demonstrated a pedal-powered airship as late as 1931. She was the daughter of a mechanic and was very mechanically inclined. In later life she was a seamstress, designing & sewing balloons and parachutes for the German Army. Two streets in Berlin are named in her honor, and a small group of enthusiasts still display some of her historical artifacts periodically at the Frankfurt International Airport.

The picture above is from a poster advertising her ascents on a 2 hour schedule from the Frankfort Zoological Gardens. The name ADLER, the backer of this flying endeavor, is still a well known manufacturer in Germany (sewing machines). There is little doubt that Katchen flew it at the exposition, but reports on her success with this vehicle have not surfaced. If it did work, it was most likely inside of an enclosed exhibition building. Since the method of mechanical propulsion was of minimal efficiency, by today's standards, would not be safe to fly outdoors.

r: J.Provan; P.van Daalen archives

K. Paulus &
The ADLER RAD

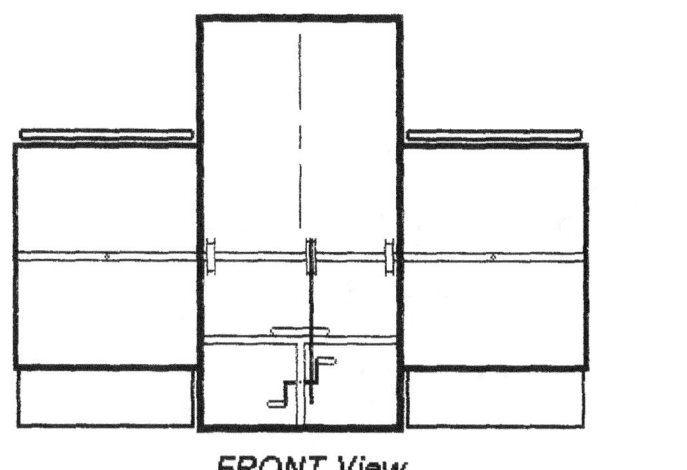

FRONT View

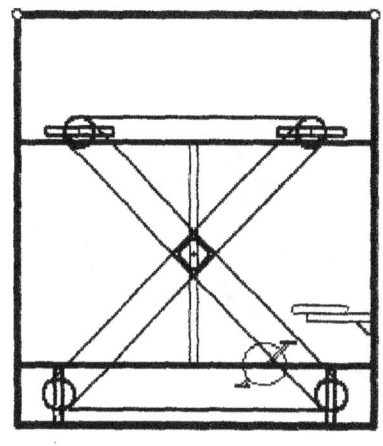

SIDE View

Pictures do not reveal the actual details of blade orientation in the rotational sequence.

Success was doubtful because of poor propulsive efficiency.

Blades must orientate for minimum drag for three quarters of the sequence.

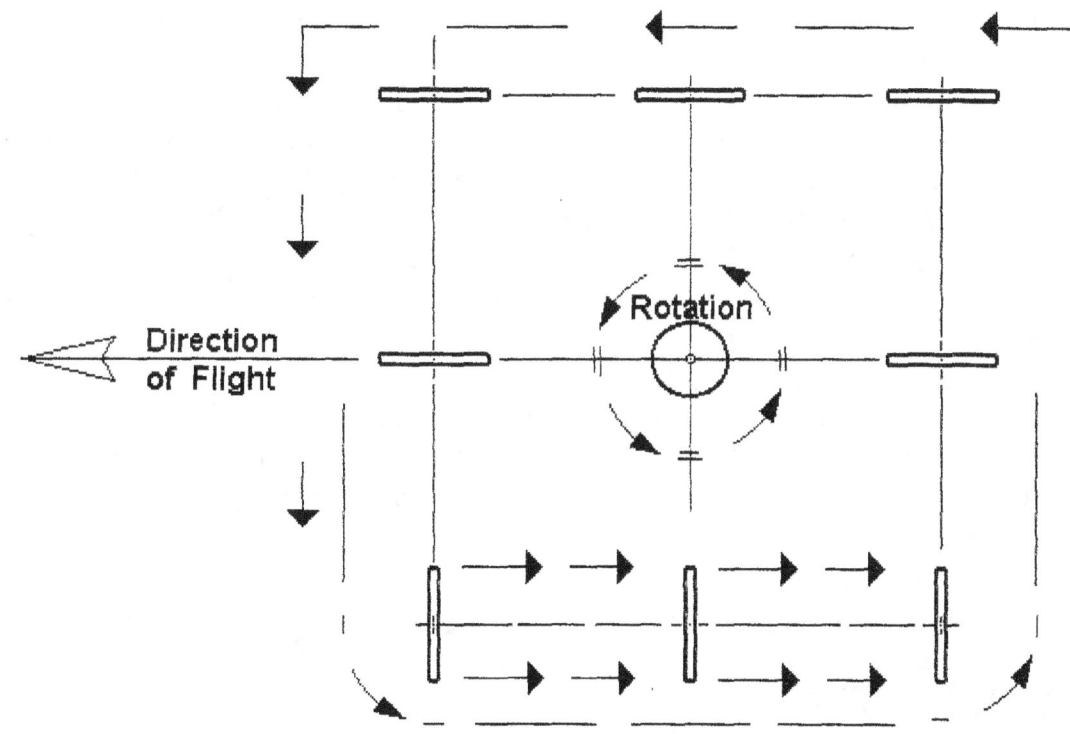

Early Muscle Power Successes

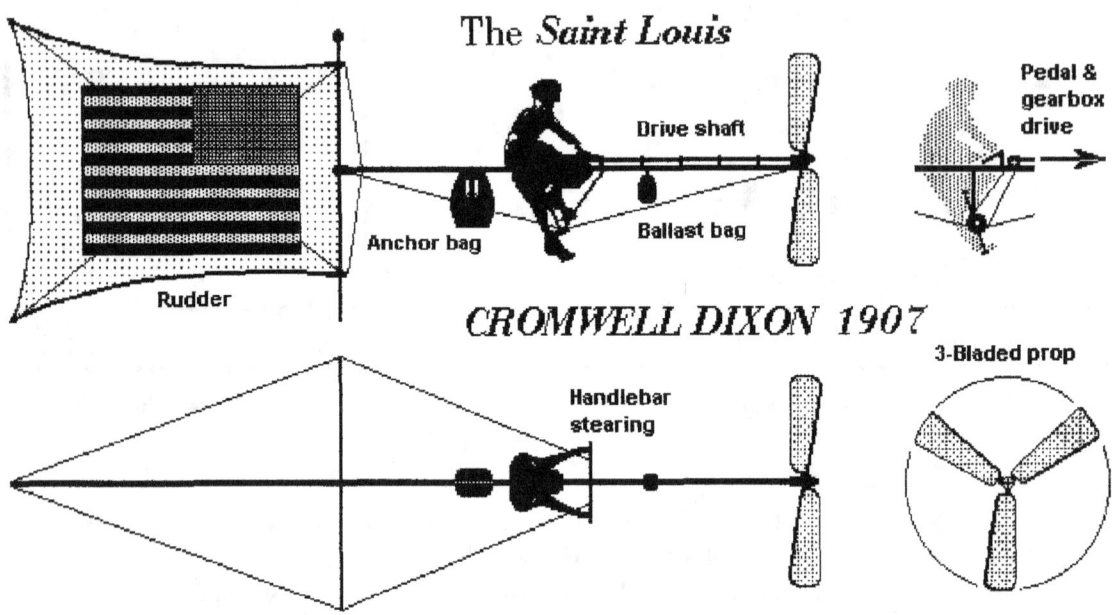

The *Saint Louis* Airship made its debut at the Worlds Fair of 1907. It turned out to be one of the few crowd pleasers of the aeronautical exhibition. The big advertising draw of the event was the collection of aeronauts and aircraft from all over the World. For various reasons, all them met with some technical failure, except the Saint Louis. It performed on cue, and thus received the lasting alocades of the media.

There were *TWO* co-builders who were also the co-aeronauts. The idea was most likely that of the then 14 year old lad, **Cromwell Dixon**, born and raised in Columbus Ohio. However, his mother whose name is lost to history, proved to be the guiding hand in design and construction. She also proved more than able in dealing with sub-contractors, and seemed to have a remarkable understanding of practical technology.

The envelope was constructed by Carl Myers (then 65 years old), who was quoted as disapproving of the envelope design, but seems to have contributed the pedal and gearbox technology. The one fascinating unit that deserves acclaim is the high efficiency, 3-bladed propeller. For low speed flight, it even has merit to this day.

Cromwell continued his love of flight, becoming wholly engrossed in powered airplanes. Unfortunately, like most early pioneers, he died in a tragic airplane crash 2 Oct. 1912 in Seattle Washington.

Icarus wasn't the last to Die Pushing His Luck

The Eagle Aloft p499-500

MUSCLE POWERED BLIMPS

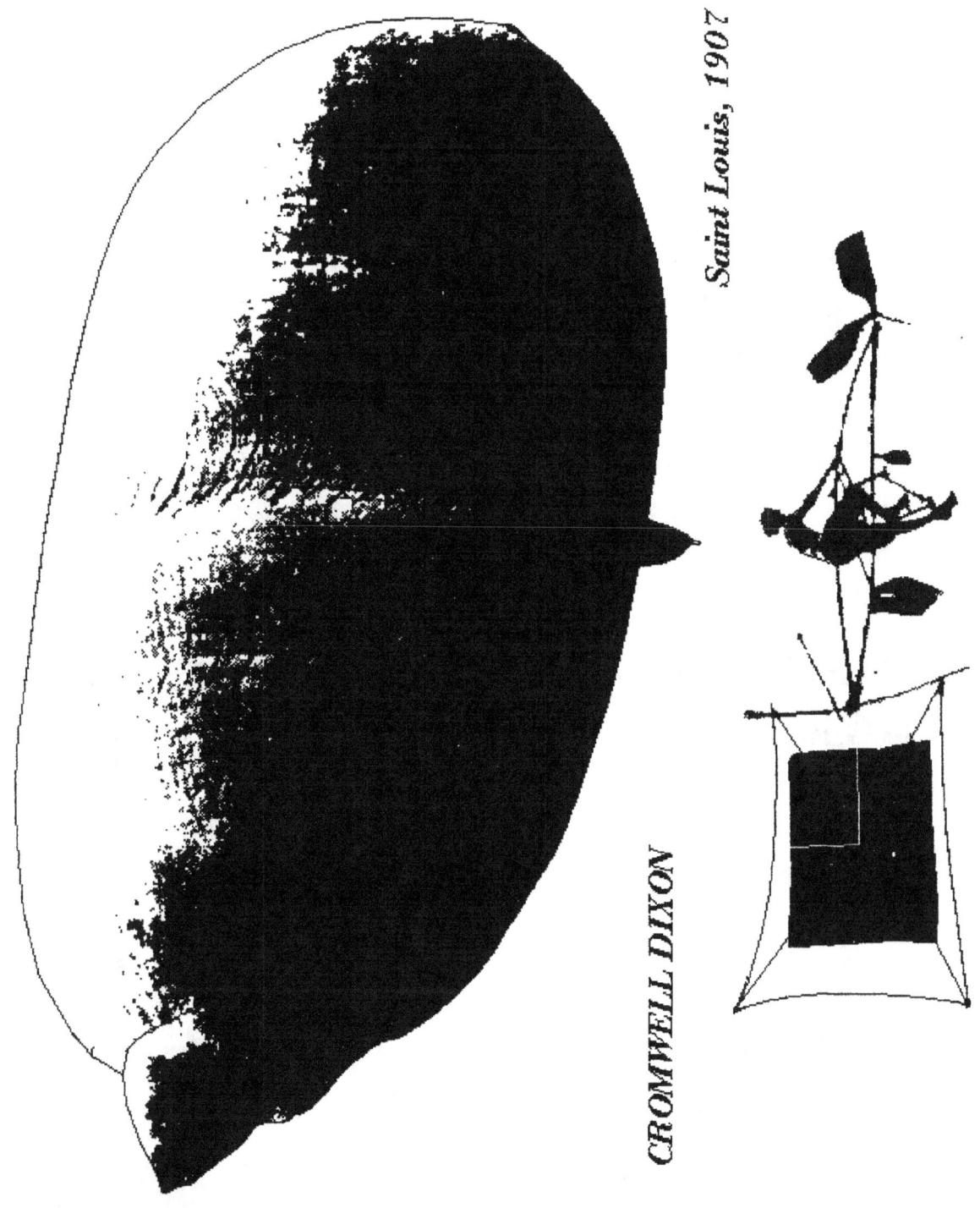

CROMWELL DIXON
Saint Louis, 1907

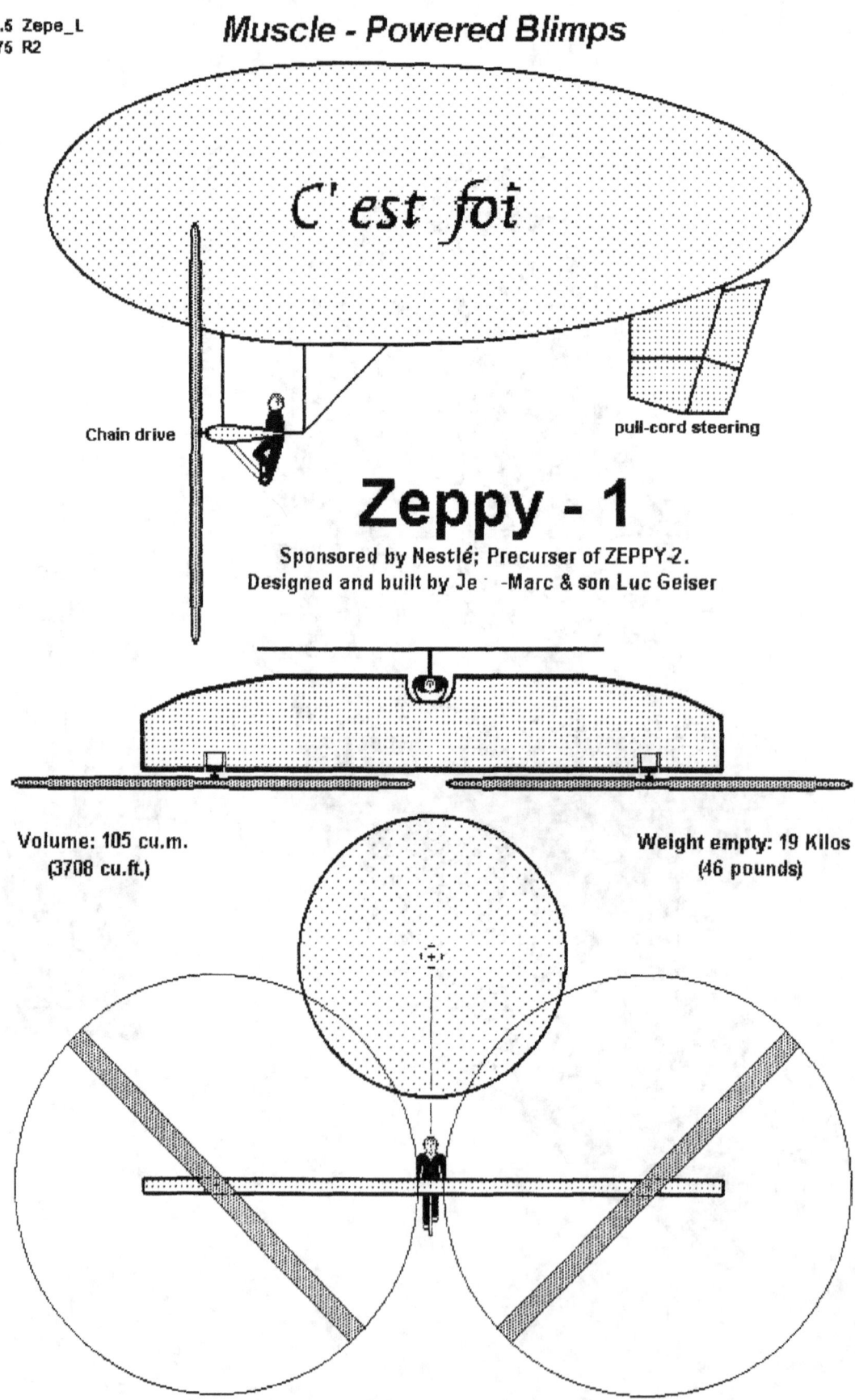

C10 ALBUQUERQUE JOURNAL Monday, February 24, 1992

Brawn Voyage

THE ASSOCIATED PRESS

A dirigible powered by human strength and sunlight is prepared for a test flight last week in Melun, south of Paris. A French aeronautical engineer and a pilot will power the aircraft from the two-seat gondola with assistance from solar power in a historic trans-Atlantic attempt from Spain to the French Indies in March.

MUSCLE POWERED BLIMPS

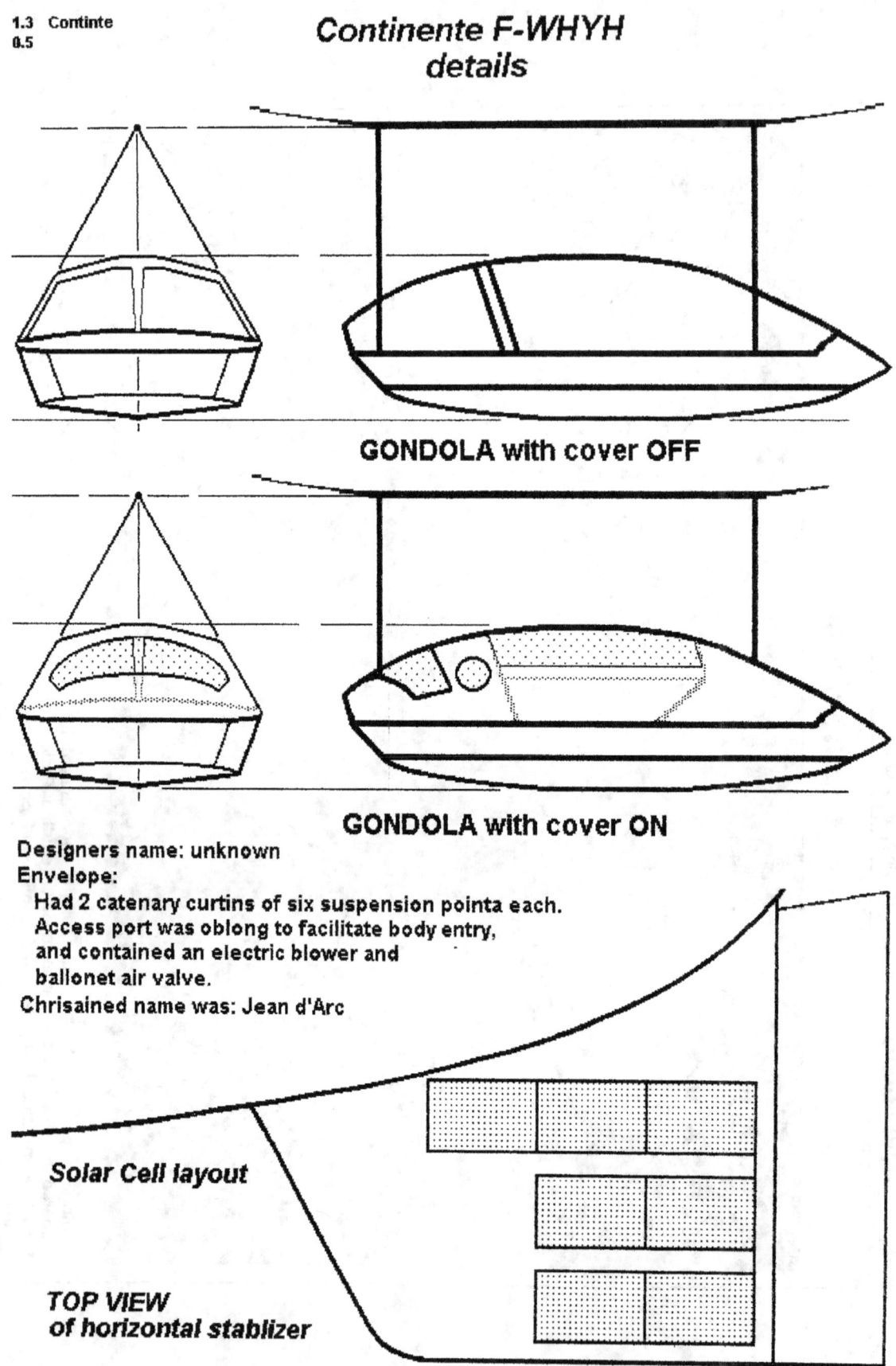

1.3 Continte
0.5

Continente F-WHYH details

GONDOLA with cover OFF

GONDOLA with cover ON

Designers name: unknown
Envelope:
   Had 2 catenary curtins of six suspension points each.
   Access port was oblong to facilitate body entry,
   and contained an electric blower and
   ballonet air valve.
Chrisained name was: Jean d'Arc

Solar Cell layout

TOP VIEW of horizontal stablizer

The Continente "Zeppy 2"
pedal-powered blimp project.

Type: 2-place, Pedal-Powered blimp; 600 cu.mts.(21,270 cu.ft) helium.
Envelope: 22m (71.5') long, 7.4m (24.27') high; Fineness ratio: 3:1.
    420 sq.mts. Polyamide fabric, with polyurethane coating.
    Weight: 86kg (190#; 3.69 oz/sq.yd.).
Suspension: Complex catenary system for distributing loads.
    Carbon keel 18m long, 15kg, connecting gondola to envelope
Propellor: Rear-mounted kevlar 4.6m (10.3') diameter 1.8kg (4 lbs),
    connected via a long chain to the pedals.
Gondola: Fibreglass, 4 sq.mts. living space, dual-place pedal system;
    Unsinkable bunks for sleeping.
Other Cooking: 30 mins per meal on small gas burner
    Water: Sea-desalinator (several centilitres per hour)
Coontrols: Pitch (up/down) via raising and lowering of nacelle, or
    with 2 water ballast tanks connected via pipe and pump.
    Direction (left/right) via classic rudder
    Guiderope: Heavy trailing line to auto-maintain 50-100m altitude;
    Hydrovane: Actually worked up to 20 km/h.(12.4 Mph).
Power: Pedal-powered with solar and hydrovane assist,
    Solar panel for recharging batteries & assist pedalling;
    Design speed 30 km/h.(18.6 Mph) maximum.
FLIGHT PLAN #1 from Sevilla (Huelva), Spain;
    Distance: 6800 km via Azores (30-35 km/h).
    Result: cancelled due to weather (missed window).
FLIGHT PLAN #2 (a year later);
    Departure: from Tenerife (Canaries).
    Destination: Venezuela.
    Estimated flight time: 8-10 days.
Improvements: microlight emergency engine installed at rear of gondola
FLIGHT LOG: Days 1 & 2 with little wind;
    Operations: Ballast taken on at midday via water scoop;
    40kg taken on (10 litres in 10 seconds);
    following increased buoyancy from solar heating of helium;
Day 3: Hydrovane breaks (tangles with guiderope);
    Electric motor burns out but airship OK.
Day 4: Dirigible advances at 25 knots; HF radio fails;
    Tropical weather front forces destination change: Brazil;
    Guiderope breaks, rendering altitude control impossible.
    Dirigible loses control and begins to free balloon;
    Crew activates emergency rip and try to detach the nacelle;
    Nacelle only unhooks at one end;
    Critical situation forces crew to jump together.
    Sudden loss of weight causes total loss of airship and videos.

Source: Marvin Johnson, Paris.

BIOGRAPHIES

CAMPBELL, Peter Carmant   USA   (1832?1890)   (HP)
e: Episcopal Institute; Stapleton Military School. p: Jeweler, Inventor. Wealthy resident of New York City. l: Author of "The Aerostat" (R629.238 A25); Builder & pilot (8 Dec. 1888) of the first successful pedal powered airship. (see C.MYERS, J.ALLEN, E. HOGAN, C.RITCHEL). m: Respected member of the Scientific community of his day; Confidant of: Samuel Morse (telegraph), and Horace Greeley (newspapers). r: The Eagle Aloft; New York Sun 15 Dec.1888; Jackson (MI) Citizen-8 Sep.1974/25.

MYERS, Carl Edgar   USA   #- 5 (1842-1925)   (GB)
See APPENDIX-C/E/H/S/AF/AR.   Aeronaut extroardinare.
Husband of balloonist Mary; Father of aeronaut Elizabeth Ariel; First name actually "Charles"; Maximum weight in lifetime was 115 pounds. b: 2 Mar.1842 Herkimer, NY. d: 30 Nov.1925, Atlanta, GA. p: Banker, Telegrapher, Photographer; Professional balloonist & Rain maker, 1870-1910 era. f: Reportedly made successful experiments W/manned kites 1854; First B-Flight 25 Aug.1880 from Sherburne, NY fairgrounds; Received B-Instruction from R.WALLISON; FAI-ACA B-License #5 issued 1908; Claimed to have carried 250,000 passengers in his lifetime (most all included were in tethered ascents); Last B-Flight (W/wife) 5 Jul.1907 (age 65); Retired 1919, moved to Atlanta (GA). l: Established balloon farm in Frankfort, NY (near Utica); Demonstrated sucessful "Sky Cycle" in 1881 (see CAMPBELL/RITCHEL); FIRST to successfully use an internal combustion engine (Curtis-Hercules) in controlled free flight (airship-2 Nov. 1903); Sold to C.BENBOW after test flight. (see WRIGHT's 15 Dec.03); Credited with the development of Linseed-oil coated B-Fabric; Made & sold first commercially available meteorological balloons in 1892. Made 21 Military Observation balloons for the U.S. Army in 1898 (mostly for meteorological purposes); Patented, made, flew, and sold many skycycles (claimed 120 pedal-powered airship ascents by 1900); Holder of many patents in Meteorology, Photography, Electricity, Metalurgy, Telegraphy, and Aviation. r: St. Nicholas-Mar.1901/p387-400; The Eagle Aloft; M.Lynch archives.

RITCHEL, Charles Francis   USA   (1850?1920)   (GB)
See APPENDIX-E/S.   Sport Balloonist, 1875-80 era.
b: Corry, PA. p: Inventor, Tinkerer, Machine maker. l: More than 150 patents. ***Epic flight attempt 22 May 1878 from Hartford, CN. Flight attempted on a self-made, pedal-powered dirigicycle weighing only 68 pounds empty. Ritchel was too heavy to fly it, but it was successfully demonstrated by a young boy (Mark QUINLAN) and a woman (Mabel Harrington), inside an exhibition hall, on 12 June1878. (see CAMPBELL Dirigicycle 1888; MYERS Skycycle 1897). r: The Eagle Aloft p491.

Muscle Power

Chapter 2

Technical Notes

Prepared by
Robert J. Recks

DESIGN CONSIDERATIONS

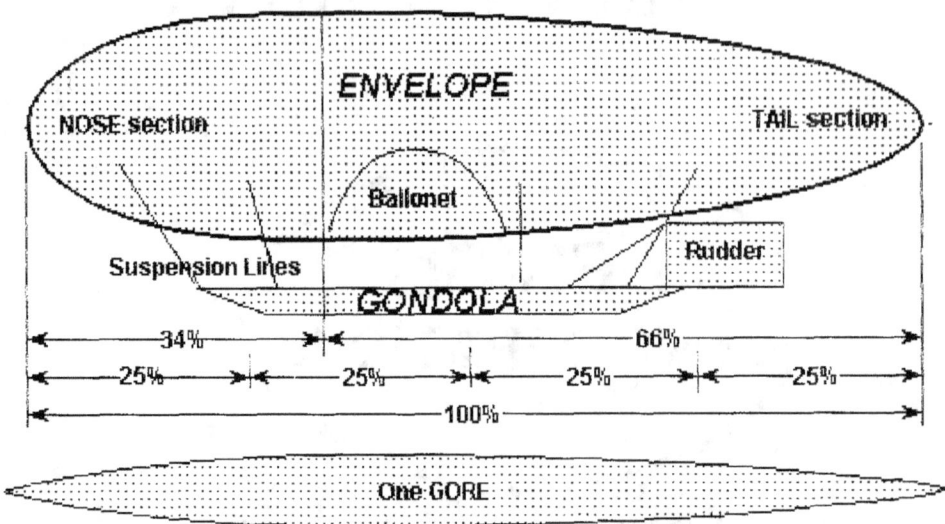

The above drawing is accompanied by standard airship nomenclature, so that we all speak the same language in this discussion.

Let me again make it clear that I am not going to design your airship for you. My intention is to introduce you to what makes a good blimp, and why. A blimp, like any other aircraft, is a series of compromises. You can make it as exotic as you can pay for it. But just because you throw money at it, doesn't mean that it will be easy to fly. And conversely, that if you made it very cheaply, doesn't mean that it won't fly nicely. There are, however, a few basic peramiters that will make it a joy to fly in relatively calm air.

The ENVELOPE (bag) is going to be your biggest time consuming expense. If (and I strongly suggest) you purchase one ready-made; you'll be better off in the long run. Assuming that you are using a single hull* envelope, here are some points to ponder:

    1. Make ALL of the other hardware items first (Gondola, Fin(s), Prop, Handling lines, Gas/Air valves, etc.), and weight them carefully _before_ you purchase the envelope. Do it the other way around, and you probably won't get off the ground. I am sad to say, 3 out of four make that mistake.

    2. Decide on how long you want it to last. One day, 1-month, 1-year, 5-years? Them pick an envelope fabric that will last that long. If you have a hangar to keep the inflated blimp in, it will last 5 times as long as if you keep it outside on a mast. If you plan to deflate it each time and store it in a box, the fabric will last only one tenth as long.

    3. Use a standard shape (Fineness Ratio) of four to one (4:1). Remember, your frontal area will be of primary importance in wind penatration. The 4:1 will give you the best air flow separation (Drag). When that 3 mph wind goes to six, you'll understand why.

NOTE: All assumptions based on a 15 mph maximum airspeed.

DESIGN CONSIDERATIONS

4. Keep your volume as small as possible (Helium is expensive). A single place blimp will require 8500 to 14,000 cubic feet of helium. The key determinant will be the overall (Gross) weight, of course. But the one item most underestimated is the operating temperature extreems. A 10 degree(F) variation during your flight will mean a 5% change in your volume. The same rule applies when you climb or descend; each change of altitude produces a corresponding change in volume. Without a ballonet to compensate, you may be approaching a pressure rupture or a saggy-bag.

5. The smoother (cleaner) the envelope surface, the less parasite drag produced. Keep the number of seams and protrusions as minamal as possible. Use the smallest mumber of gores as practical. Do not even consider using multi-panel gores on a small blimp.

6. Use a Double Sphereoid profile (shape) for simplicity. Make the Nose parabola 34%, and the Tail parabola 66% of the inflated length. It is the easiest to calculate, and closely resembles the most efficient (air flow) design.

7. Calculate your gore coordinates in 100 incraments (100%). Since the first & last 20% will have the greatest curvature change rate, plot at those extream, widths at 2% incraments. The rest can be done in 5% incraments.

8. An emergency deflation system must be incorporated capable of ground incapacitation (100 pounds heavy) in 10 seconds (or less) with a 50 pound pull device (cord) actuation.

*For simplicity, I don't recommend a Double Hull envelope (2 independant cells, one for the gas bladder, and one for the bladder container). But if you do, make the outer of a stretch resistant fabric (polyester), and the inner (usually polyethelene) at least 10% larger.

The **GONDOLA** (car) will be the most complex project, and require the most thought. It is entirely possible that you will even call the first one a "prototype". Comfort will always take precidence over utility; but don't go overboard. You have to incorporate many subsystems (controls), all doing a specific job, and all within the pilots reach. For the pilots maximum efficiency, consider:

1. The gondola structure should be absolutely rigid to preclude loss of control from flexing, but able to sustain minor brusing of landing or handling. Welded chrome-molly tubing seems to work best, aluminum has been used successfully, composit hasn't been tried as of this writing.

2. The pilots position for efficient pedaling. More than 10 have been tried, but the best seems to be a 30% recline. It does not allow maximum force on the pedals (as you tend to slide back on the seat), nor does it reduce the profile drag. It is just a good compromise. (It isn't condusive to a relaxing pee enroute either)

3. A (bicycle) chain drive to the prop is about the simplest to understand and to route around obstructions. Do not consider using plastic chain. Stick with common bicycle chain off the roll. Titanium is light and available, but at 10 times the cost.

4. The prop mast position must be easily adjustable by the pilot. It should have a detent range from -30 to +30 degrees from centerline. Assuming a 20 pound heavy takeoff, the cruise detent will about 5 degrees up.

NOTE: All assumptions based on a 15 mph maximum airspeed.

DESIGN CONSIDERATIONS

5. The steering (Rudder) cables should always be under tension for smooth control. Bungee return pull cords take more arm muscle, but are sufficient in no-wind conditions. A ratchet crank will give a more positive rudder force but requires the installation of a rear view mirror to verify position. The crank will be required at higher airspeeds.

6. Provisions should be made for water ballast container(s); but must have provisions for both micro (tweeking) adjustment AND rapid disposal (emergency, red handle). A motorhome water gate valve works best.

7. If you are using a ballonet to control the envelope pressurization, you will need a small rotary (or bellows) air pump that can be engaged in the chain drive. A small relief valve, with a predetermined (1"WP) setting, is installed in the appendix line.

8. The rudder obviously needs to be light and strong. Here some innovation and composits can be taken advantage of. Hard-dense foams are available in sheets that make ideal contoured airfoil ribs. New adhesive coverings are simple to use, and make beautiful seamless skins. As a starting point, at least 36 square feet of rudder surface will be required for 50' long blimp. Offset hinge points can be used very effectively at highrt airspeeds.

9. Suspension lines from the envelope should be capable of 6 times the gondola normal loaded weight. One sixteenth inch stranded steel cable is sufficient. Spectra cord is also a good alternative but special attention must be given to securing the ends to preclude slippage.

Now if that isn't enough, the gondola should be dismantable into managable lengths for transport. Plan your gondola with the transport vehicle in mind, or vise versa.

NOTE: All assumptions based on a 15 mph maximum airspeed.

AERODYNAMIC PROFILE

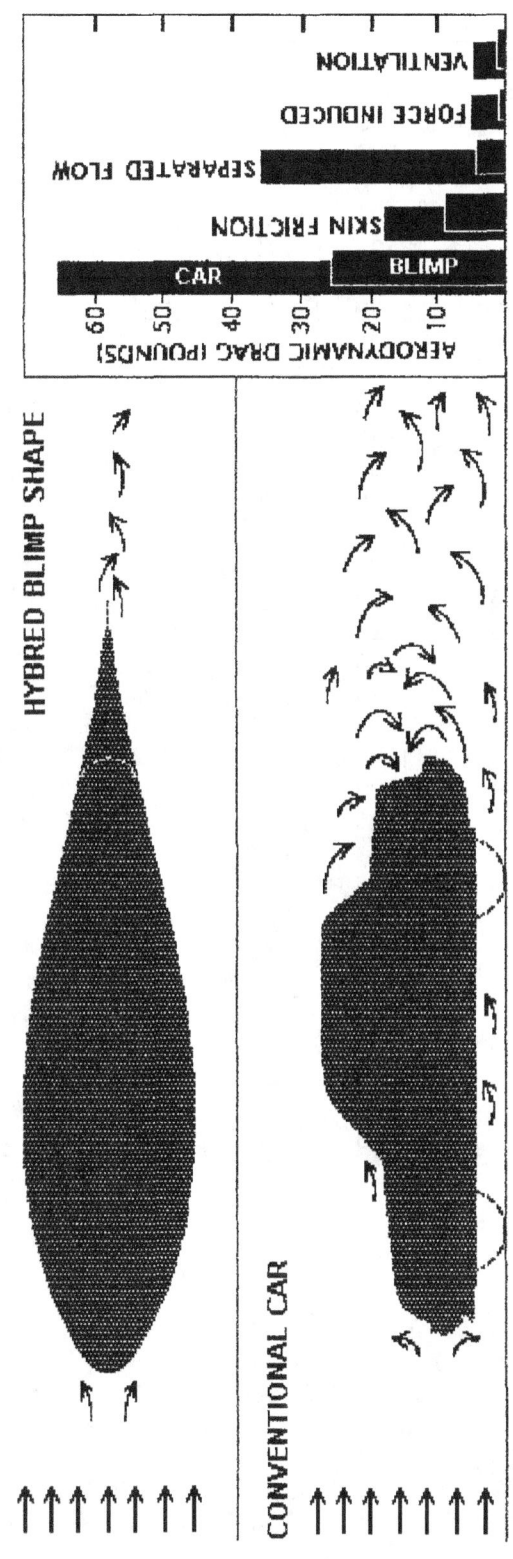

AERODYNAMIC PROFILE (contour) lets *a blimp* slip through the air, leaving little disturbed flow in its wake. (Flow lines are shown for the center plane only.) The graph shows the aerodynamic drag from four types of flow for a vehicle speed of 55 mph. Aerodynamic drag is proportional to speed squared; the power consumed, drag times the speed, is proportional to speed cubed. The *blimp* expends 1.84 horsepower to overcome the drag. A conventional car expends 9.52 hp. The front third of the *blimp* has a smooth boundary layer and contributes less than 10 percent of the total skin friction. On a conventional car the boundary layer becomes turbulent at the sharp layer and grows rapidly. Separated flow breaks away from the object and forms eddies, which mix with the outer Sow. For the *blimp*, separated flow behind the wheels and wheel wells causes about a third of the total drag. The much larger drag for a conventional car comes mainly from separated how in the rear and behind the heels. Induced drag results when local downward flow lifts the vehicle. The blimp has no aerodynamic lift and so has no induced drag. At 55 mph a conventionsal car would experience 50 poounds of lift and five pounds of induced drag. Internal ventilating flows are also a source of drag.

MUSCLE POWERED BLIMPS

VOLUMES - Aspect Ratio 4:1

Length	Dia.	Radius	R x R	x 2.0944	Cu.Feet	Cu.Mt.	Lbs.Lift
1	0.25	0.125	0.015625	0.032725	0.032725	0.0009267	0.002029
2	0.5	0.25	0.0625	0.1309	0.2618	0.0074133	0.0175406
3	0.75	0.375	0.140625	0.294525	0.883575	0.0250201	0.0591995
4	1	0.5	0.25	0.5236	2.0944	0.0593068	0.1403248
5	1.25	0.625	0.390625	0.818125	4.090625	0.1158336	0.2740719
6	1.5	0.75	0.5625	1.1781	7.0686	0.2001604	0.4735962
7	1.75	0.875	0.765625	1.603525	11.224675	0.3178474	0.7520532
8	2	1	1	2.0944	16.7552	0.4744544	1.1225984
9	2.25	1.125	1.265625	2.650725	23.856525	0.6755415	1.5983872
10	2.5	1.25	1.5625	3.2725	32.725	0.9266687	2.192575
11	2.75	1.375	1.890625	3.959725	43.556975	1.2333961	2.9183173
12	3	1.5	2.25	4.7124	56.5488	1.6012835	3.7887696
13	3.25	1.625	2.640625	5.530525	71.896825	2.0358912	4.8170873
14	3.5	1.75	3.0625	6.4141	89.7974	2.542779	6.0164258
15	3.75	1.875	3.515625	7.363125	110.44688	3.1275069	7.3999406
16	4	2	4	8.3776	134.0416	3.7956351	8.9807872
17	4.25	2.125	4.515625	9.457525	160.77793	4.5527234	10.772121
18	4.5	2.25	5.0625	10.6029	190.8522	5.404332	12.787097
19	4.75	2.375	5.640625	11.813725	224.46078	6.3560207	15.038872
20	5	2.5	6.25	13.09	261.8	7.4133497	17.5406
21	5.25	2.625	6.890625	14.431725	303.06623	8.581879	20.305437
22	5.5	2.75	7.5625	15.8389	348.4558	9.8671685	23.346539
23	5.75	2.875	8.265625	17.311525	398.16508	11.274778	26.67706
24	6	3	9	18.8496	452.3904	12.810268	30.310157
25	6.25	3.125	9.765625	20.453125	511.32813	14.479199	34.258984
26	6.5	3.25	10.5625	22.1221	575.1746	16.287129	38.536698
27	6.75	3.375	11.390625	23.856525	644.12618	18.23962	43.156454
28	7	3.5	12.25	25.6564	718.3792	20.342232	48.131406
29	7.25	3.625	13.140625	27.521725	798.13003	22.600523	53.474712
30	7.5	3.75	14.0625	29.4525	883.575	25.020055	59.199525
31	7.75	3.875	15.015625	31.448725	974.91048	27.606388	65.319002
32	8	4	16	33.5104	1072.3328	30.365081	71.846298
33	8.25	4.125	17.015625	35.637525	1176.0383	33.301694	78.794568
34	8.5	4.25	18.0625	37.8301	1286.2234	36.421787	86.176968
35	8.75	4.375	19.140625	40.088125	1403.0844	39.730921	94.006653
36	9	4.5	20.25	42.4116	1526.8176	43.234656	102.29678
37	9.25	4.625	21.390625	44.800525	1657.6194	46.938551	111.0605
38	9.5	4.75	22.5625	47.2549	1795.6862	50.848166	120.31098
39	9.75	4.875	23.765625	49.774725	1941.2143	54.969062	130.06136
40	10	5	25	52.36	2094.4	59.306798	140.3248
41	10.25	5.125	26.265625	55.010725	2255.4397	63.866935	151.11446
42	10.5	5.25	27.5625	57.7269	2424.5298	68.655032	162.4435
43	10.75	5.375	28.890625	60.508525	2601.8666	73.67665	174.32506
44	11	5.5	30.25	63.3556	2787.6464	78.937348	186.77231
45	11.25	5.625	31.640625	66.268125	2982.0656	84.442687	199.7984
46	11.5	5.75	33.0625	69.2461	3185.3206	90.198226	213.41648
47	11.75	5.875	34.515625	72.289525	3397.6077	96.209526	227.63971
48	12	6	36	75.3984	3619.1232	102.48215	242.48125
49	12.25	6.125	37.515625	78.572725	3850.0635	109.02165	257.95426
50	12.5	6.25	39.0625	81.8125	4090.625	115.83359	274.07188

MUSCLE POWERED BLIMPS

VOLUMES - Aspect Ratio 4:1

L	D	R	R x R	x 2.0944	Cu.Ft.	Cu.Mt.	Lbs.Lift
51	12.75	6.375	40.640625	85.117725	4341.004	122.92353	290.84727
52	13	6.5	42.25	88.4884	4601.3968	130.29704	308.29359
53	13.25	6.625	43.890625	91.924525	4871.9998	137.95966	326.42399
54	13.5	6.75	45.5625	95.4261	5153.0094	145.91696	345.25163
55	13.75	6.875	47.265625	98.993125	5444.6219	154.17451	364.78967
56	14	7	49	102.6256	5747.0336	162.73785	385.05125
57	14.25	7.125	50.765625	106.32353	6060.4409	171.61256	406.04954
58	14.5	7.25	52.5625	110.0869	6385.0402	180.80419	427.79769
59	14.75	7.375	54.390625	113.91573	6721.0278	190.31829	450.30886
60	15	7.5	56.25	117.81	7068.6	200.16044	473.5962
61	15.25	7.625	58.140625	121.76973	7427.9532	210.33619	497.67287
62	15.5	7.75	60.0625	125.7949	7799.2838	220.8511	522.55201
63	15.75	7.875	62.015625	129.88553	8182.7881	231.71073	548.2468
64	16	8	64	134.0416	8578.6624	242.92064	574.77038
65	16.25	8.125	66.015625	138.26313	8987.1031	254.4864	602.13591
66	16.5	8.25	68.0625	142.5501	9408.3066	266.41355	630.35654
67	16.75	8.375	70.140625	146.90253	9842.4692	278.70766	659.44543
68	17	8.5	72.25	151.3204	10289.787	291.3743	689.41574
69	17.25	8.625	74.390625	155.80373	10750.457	304.41901	720.28062
70	17.5	8.75	76.5625	160.3525	11224.675	317.84737	752.05323
71	17.75	8.875	78.765625	164.96673	11712.637	331.66493	784.74671
72	18	9	81	169.6464	12214.541	345.87725	818.37423
73	18.25	9.125	83.265625	174.39153	12730.581	360.48988	852.94895
74	18.5	9.25	85.5625	179.2021	13260.955	375.5084	888.48401
75	18.75	9.375	87.890625	184.07813	13805.859	390.93837	924.99258
76	19	9.5	90.25	189.0196	14365.49	406.78533	962.4878
77	19.25	9.625	92.640625	194.02653	14940.042	423.05485	1000.9828
78	19.5	9.75	95.0625	199.0989	15529.714	439.75249	1040.4909
79	19.75	9.875	97.515625	204.23673	16134.701	456.88382	1081.025
80	20	10	100	209.44	16755.2	474.45438	1122.5984
81	20.25	10.125	102.51563	214.70873	17391.407	492.46975	1165.2243
82	20.5	10.25	105.0625	220.0429	18043.518	510.93548	1208.9157
83	20.75	10.375	107.64063	225.44253	18711.73	529.85713	1253.6859
84	21	10.5	110.25	230.9076	19396.238	549.24026	1299.548
85	21.25	10.625	112.89063	236.43813	20097.241	569.09043	1346.5151
86	21.5	10.75	115.5625	242.0341	20814.933	589.4132	1394.6005
87	21.75	10.875	118.26563	247.69553	21549.511	610.21413	1443.8172
88	22	11	121	253.4224	22301.171	631.49879	1494.1785
89	22.25	11.125	123.76563	259.21473	23070.111	653.27272	1545.6974
90	22.5	11.25	126.5625	265.0725	23856.525	675.5415	1598.3872
91	22.75	11.375	129.39063	270.99573	24660.611	698.31067	1652.2609
92	23	11.5	132.25	276.9844	25482.565	721.58581	1707.3318
93	23.25	11.625	135.14063	283.03853	26322.583	745.37247	1763.613
94	23.5	11.75	138.0625	289.1581	27180.861	769.67621	1821.1177
95	23.75	11.875	141.01563	295.34313	28057.597	794.50259	1879.859
96	24	12	144	301.5936	28952.986	819.85718	1939.85
97	24.25	12.125	147.01563	307.90953	29867.224	845.74552	2001.104
98	24.5	12.25	150.0625	314.2909	30800.508	872.17318	2063.634
99	24.75	12.375	153.14063	320.73773	31753.035	899.14573	2127.4533
100	25	12.5	156.25	327.25	32725	926.66872	2192.575

METRIC CONVERSION CHART

To Convert... U.S. System	To... Metric System	Multiply by...	To Convert... Metric System	To... U.S. System	Multiply by...
Density					
lb/in³	kg/m³	27,680	kg/m³	lb/in³	0.000036
lb/ft³	g/cm³	0.0160	g/cm³	lb/ft³	62.43
lb/ft³	kg/m³	16.0185	kg/m³	lb/ft³	0.0624
lb/in³	g/cm³	27.68	g/cm³	lb/in³	0.03613
Temperature					
in/(in · °F)	m/(m · °C)	1.8	m/(m · °C)	in/(in · °F)	0.556
°F	°C	(°F − 32)/(1.8)	°C	°F	1.8°C + 32
°F	K	(°F + 459.67)/(1.8)	K	°F	1.8K − 459.67
Pressure					
psi	kPa	6.8948	kPa	psi	0.145
psi	MPa	0.00689	MPa	psi	145
psi	GPa	0.00000689	GPa	psi	145,038
psi	bar	0.0689	bar	psi	14.51
Energy and Power					
ft · lbf	J	1.3558	J	ft · lbf	0.7376
in · lbf	J	0.113	J	in · lbf	8.850
ft · lbf/inch	J/m	53.4	J/m	ft · lbf/inch	0.0187
ft · lbf/inch	J/cm	0.534	J/cm	ft · lbf/inch	1.87
ft · lbf/in²	kJ/m²	2.103	kJ/m²	ft · lbf/in²	0.4755
kW	metric horsepower	1.3596	metric horsepower	kW	0.7355
U.S. horsepower	kW	0.7457	kW	U.S. horsepower	1.3419
Btu[a]	J	1055.1	J	Btu[a]	0.00095
Btu[a]	W · h	0.2931	W · h	Btu[a]	3.412
Btu[a] · in/ (h · ft² · °F)	W/(m · K)	0.1442	W/(m · K)	Btu[a] · in/ (h · ft² · °F)	6.933
Btu[a]/lb	kJ/kg	2.326	kJ/kg	Btu[a]/lb	0.4299
Btu[a]/(lb · °F)	J/(kg · °C)	4187	J/(kg · °C)	Btu[a]/(lb °F)	0.000239
V/mil	MV/m	0.0394	MV/m	V/mil	25.4
Output					
lb/min	g/s	7.560	g/s	lb/min	0.1323
lb/h	kg/h	0.4536	kg/h	lb/h	2.2046
Velocity					
in/min	cm/s	0.0423	cm/s	in/min	23.6220
ft/s	m/s	0.3048	m/s	ft/s	3.2808
Viscosity					
poise	Pa · s	0.1	Pa · s	poise	10

[a] International table

METRIC CONVERSION CHART

To Convert... U.S. System	To... Metric System	Multiply by...	To Convert... Metric System	To... U.S. System	Multiply by...
Length					
mil	millimeter	0.0254	millimeter	mil	39.37
inch	millimeter	25.4	millimeter	inch	0.0394
inch	centimeter	2.54	centimeter	inch	0.3937
foot	centimeter	30.48	centimeter	foot	0.0328
foot	meter	0.3048	meter	foot	3.2808
yard	meter	0.9144	meter	yard	1.0936
Area					
inch^2	millimeter^2	645.16	millimeter^2	inch^2	0.0016
inch^2	centimeter^2	6.4516	centimeter^2	inch^2	0.155
foot^2	centimeter^2	929.03	centimeter^2	foot^2	0.0011
foot^2	meter^2	0.0929	meter^2	foot^2	10.7639
yard^2	meter^2	0.8361	meter^2	yard^2	1.1960
Volume, Capacity					
inch^3	centimeter^3	16.3871	centimeter^3	inch^3	0.061
fluid ounce	centimeter^3	29.5735	centimeter^3	fluid ounce	0.0338
quart (liquid)	decimeter^3 (liter)	0.9464	decimeter^3 (liter)	quart (liquid)	1.0567
gallon (U.S.)	decimeter^3 (liter)	3.7854	decimeter^3 (liter)	gallon (U.S.)	0.2642
gallon (U.S.)	meter^3	0.0038	meter^3	gallon (U.S.)	264.17
foot^3	decimeter^3	28.3169	decimeter^3	foot^3	0.0353
foot^3	meter^3	0.0283	meter^3	foot^3	35.3147
yard^3	meter^3	0.7646	meter^3	yard^3	1.3079
in^3/lb	m^3/kg	0.000036	m^3/kg	in^3/lb	27,680
ft^3/lb	m^3/kg	0.0624	m^3/kg	ft^3/lb	16.018
Mass					
ounce (avdp.)	gram	28.3495	gram	ounce (avdp.)	0.03527
pound	gram	453.5924	gram	pound	0.0022
pound	kilogram	0.4536	kilogram	pound	2.2046
pound	metric ton	0.00045	metric ton	pound	2204.6
U.S. ton (short)	metric ton	0.9072	metric ton	U.S. ton (short)	1.1023
Force					
lbf	N	4.448	N	lbf	0.225

Standard Metric Symbols

Symbol	Term	Symbol	Term
A	ampere	kg	kilogram
bar	bar	L	liter
cd	candela	m	meter
C	celsius[a]	N	newton
g	gram	Pa	pascal
h	hour	S	siemens
Hz	hertz	s	second
J	joule	t	metric ton
K	kelvin	V	volt
		W	watt

[a]Formerly called Centigrade

[b]These prefixes may be used with all metric units

Metric Prefixes[b]

Numerical Value	Term	Symbol
10	deca	da
10^2	hecto	h
10^3	kilo	k
10^6	mega	M
10^9	giga	G
10^{12}	tera	T
10^{-1}	deci	d
10^{-2}	centi	c
10^{-3}	milli	m
10^{-6}	micro	μ
10^{-9}	nano	n
10^{-12}	pico	p

RECORDS

Introduction

Competition has certainly been a part of the story of mankind since the beginning. The Olympic games date from at least to 776 B.C. and probably as far back to 1370 B.C. They could hardly have sprung full-grown without some earlier attempts at pitting one contestant against another.

Just as a long time must have elapsed between man learning to walk upright and his then inventing running races, there was a long time between the first manned flight in a balloon in 1783 and the first serious balloon Competition in the final years of the 19th century.

But the invention of the airplane in 1903 was followed quickly by the first official speed and altitude records in 1906 and the first formal air races in 1909. It was undoubtedly the formation of the Federation Aeronautique Interntional (FAI) in 1905 that precipitated the rush to record achievements and to encourage better and faster and further and higher.

The first official recognition of aviation records was by the Aero Club of France in 1906. An adventurous Brazilian, Alberto Santos-Dumont, ventured aloft on November 12, 1906, at Bagatelle, France, for one of the first successful flying sessions in Europe. When he was finished for the day he had became the first official holder of the three most important records.

Speed 25.66 mph   Distance 721.8 feet   Duration 21.2 seconds.

The FAI granted international record status to these and other French records at its meeting of October 18, 1910, in Paris. From then on, the connection between the national aero club which was responsible for supervising all domestic record attempts, and the FAI which added its blessing to these it deemed worthy, has been a constant and most successful one.

The procedure followed by the National Aeronautic Association (NAA) in the United States progresses from the applicant filing a sanction request for one or more record categories, to the granting of a limited, but exclusive sanction, to the speed, distance. altitude or other record attempt supervised in the held by NAA-approved specialists. Once the attempt has been completed and the existing record surpassed (by a significant amount), the detailed paperwork is forwarded to NAA Headquarters in Washington, D.C., where it is reviewed and, in most cases, approved as a National Record. The resultant record dossier is then sent to FAI Headquarters in Paris for consideration as a World record.

A major revision of the Category & Class divisions were made in 1987, separating BA (engine powered) from BI (muscle powered) airships. As of this writing*, NO records have been applied for in the BI-Category

Source: tional Aeronautic Assoc. *10/22/98

MUSCLE POWERED BLIMPS

RECORDS

The record books grew and grew. A couple of dozen record categories became hundreds and then thousands. New categories replaced old, out-dated ones, adding new opportunities for ambitious airmen. The latest revision was in 1987 that also recognized the new potential of "Muscle" power. The achievements of the Kremer Prize could not be ignored.

At this writing, NO Class-61 (muscle power) records exist. ALL of Brian Allen's records are for BA (powered) class without regard to what kind of power. ***Therefore, the new BI-Class records will go to the FIRST pilot that applies.***

However, I think it only fair to use these as benchmarks in starting; But only volumes less than 14,000 cu.ft. are considered:

ALTITUDE:
Class BA1   None
Class BA2   1898 meters (6227') Hokan Colting   Canada   11/08/92

DISTANCE:
Class BA1   94.86 Km. (59 Sm.) Don Cameron   U.K.   10/12/90
Class BA2   same

DURATION:
Class BA1   8:50:12   Brian Allen   USA   2/12/85
Class BA2   same

SPEED: None

For help in establishing a new record contact me or:

The AIRSHIP Committee
BALLOON FEDERATION of AMERICA
P.O. Box 400
Indianola, IA 50125

Records Division
NATIONAL AERONAUTIC ASSOCIATION
1815 N.Ft. Meyer Drive #700
Arlington, VA 22209

FEDERATION AERONAUTIQUE INTERNATIONALE
8 Rue Galilée (Cedex 16)
Paris 75781 FRANCE

Source: National Aeronautic Assoc.

Muscle Power

Details and Compromises

Assembled by
Robert J. Recks

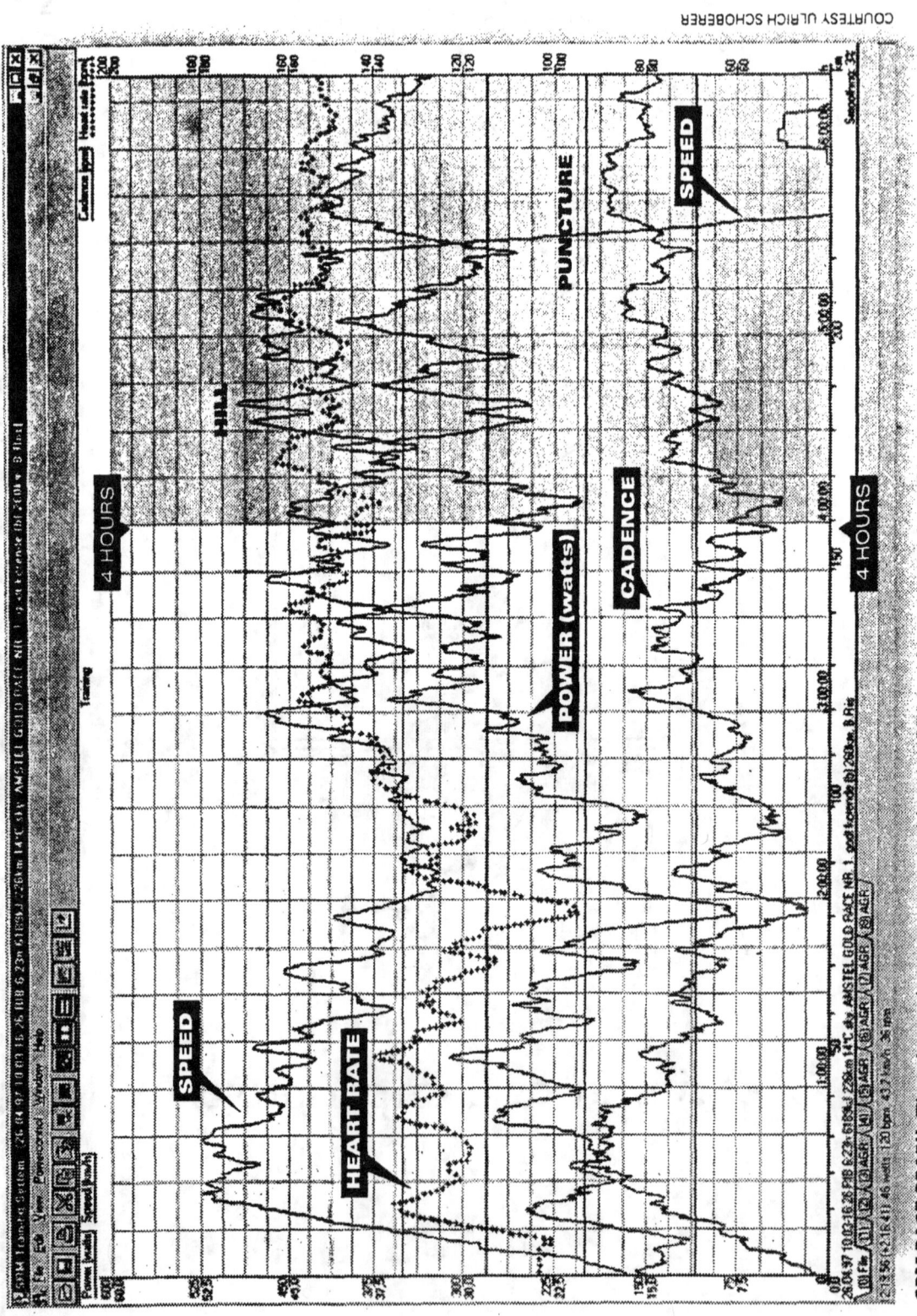

SRM RACE GRAPH: The chart above, generated by the SRM software, displays Bjarne Riis's power, heart rate, cadence and speed for the 1997 Amstel Gold Race.

COURTESY ULRICH SCHOBERER

MUSCLE POWERED BLIMPS

Pertinent Notes to FAI Sporting Code

Muscle Powered Airships still have the option of competing in either or both of two classes. Either in the BA (motorized) and/or BI (muscle powered). The class size volumes are the same for all lighter-than-air aircraft.

Sub-Class English Equivalents:

Class	Volume(cu.mt.)	Volume(cu.ft)	lift @ 0.61 #/cu.ft.
B?1	up to 400 =	14,126	862 pounds
B?2	401 - 900 =	31,783	1939 "
B?3	901 - 1,600 =	56,503	3447 "

NOTE 1: Volume is that enclosed of the spheroid envelope only, AND does *NOT* include any inflated stabilizing surfaces.

NOTE 2: The ONLY Gas airship speed record at this time* is in a 42.000 cu.ft (BA-3 class) envelope. ALL Records for Class "BI" are available for the first one that tries.

NOTE 3: It is impossible to beat this record because the aerodynamic drag requires more horsepower than a human body can produce. Speed is a logarithmic function. To double the speed, you must quadruple the horsepower.

NOTE 4: Since aerodynamic drag is the single most limit to muscle power, only class 1 or 2 will be considered as practical.

Calculating Volumes: (ref. S/P 2.2.2)

Pull the Bow Apex from the Stern Apex under 10 lbs. of tension;
   Let G = the gore length, and **L = .92298 x G**
   Let L = the inflated distance from the bow to the stern;
Pull at the Greatest Diameter with 10 lbs. of tension;
   Double that dimension to get the circumference (C);
   Diameter (D) = C / 3.14159
To determine volume (V), use the formula: **V = .50564 x D squared x L**

NOTE 4: As of this date*, declared manufacturers volumes have never been challenged;
   One Cubic Meter equals 35.3147 Cubic Feet (allowed error).

Speed Records:

These calculations will have to be extended as technology advances;
To beat the previous record over a 3 Kilometer (straight line) course by 3%:

Speed		Time	
Kph.	3% faster	Duration	3% faster
10	10.3	18 minutes	17:27.6
11	11.33	16.364	15:52.3
12	12.36	15	14:33.0
13	13.39	13.846	13:25.9
14	14.42	12.857	12:28.3
15	15.45	12	11:38.4

NOTE 5: One Kilometer = 9842.4 feet, or 1.86 Statute Miles.

*10 Oct.98

MUSCLE POWERED BLIMPS

HELIUM LIFT / ALTITUDE TABLE
(international standard atmosphere)

ALTITUDE in 1000's feet	PRESSURE in Inches	Milabars	TEMPERATURE in F-Degrees	R-Degrees	C-Degrees	HELIUM Lift/Expansion cubic feet	coefficient
0	29.92	1013.25	59.01	518.7	15.01	0.065949	1.001
1	28.86	977.17	55.43	515.1	13.02	0.064043	1.028
2	27.82	942.13	51.86	511.5	11.04	0.062177	1.061
3	26.82	902.12	48.31	508.1	9.06	0.060351	1.093
4	25.84	875.11	44.74	504.4	7.08	0.058569	1.126
5	24.91	843.07	41.17	500.8	5.09	0.056828	1.161
6	23.89	811.99	37.61	497.3	3.11	0.055127	1.196
7	23.09	781.85	34.04	493.7	1.13	0.053458	1.234
8	22.23	752.62	30.47	490.2	-0.85	0.051836	1.272
9	21.39	742.28	26.91	486.6	-2.83	0.050253	1.312
10	20.58	696.81	23.34	483.1	-4.81	0.048703	1.354

Constants: (with all other factors constant)
1. The lift of an airship varies with a change in volume.
2. The lift of an airship varies with a change in barometric pressure.
3. The lift of an airship varies with a change in temperature.
4. The lift of an airship varies with a change in humidity.
5. The barometric pressure decreases approximately 1-inch for each 1000' of altitude.
6. The temperature will decrease approximatly 1-degree of Fahrenheit for each 300 feet ascent.

Increases:
2. The lift of gas increases as barometric pressure increases,
    and decreases if the pressure decreases.
3. The lift of a fixed volume of gas decreases if the atmospheric temperature increases,
    and increases if the temperature decreases.
4. The lift does not change due to a change in barometric pressure
    if the gas is free to expand.
5. The lift decreases as the Atmospheric Humidity increases,
    for a fixed volume of gas
6. The lift does not change in lift when air and gas temperature change an equal amount,
    if the gas is free to expand.
7. An airship in equilibrium at any altitude will be in equilibrium at the surface,
    providing there is no superheating of the gas.
8. An airship rising from the surface in equilibriun will be in equilibrium at any ultitude
    below pressure height, providing no weight is lost and there is no superheating.

Numbers: (if the gas is free to expand)
13. The gas volume will increase approximately 1% for each 375 feet of altitude.
11. The gas volume will increase 1% for every 5 F-degrees of temperature.
12. The lift is increased 1% for every 5 F-degrees of superheat.

MUSCLE POWERED BLIMPS

AIRSHIP MATHEMATICS
Predicting Speed
(Prandtl Formula*)

Problem 1: What speed can a 195,000 cubic foot airship, with two engines totaling 300 horsepower, expect to achieve?

Formula: $$v = \left(\frac{HPt \times HP_1 \times E \times F}{Cd \times p \times k} \right)^{.35} = \text{velocity in fps}$$

Where:
- HPt = Total Horse Power available = 300
- HP_1 = One Horse Power = 550
- E = Propeller Efficiency (60%) = .60
- F = Gondola Resistance (40%) = .40
- Cd = Envelope shape coefficient = .0136
- p = Atmospheric Density (S.L.) = .00237
- k = (vol.) = (195,000 coefficient) = 3376.4

Setup:
$$v = \left(\frac{300 \times 550 \times .60 \times .40}{.0136 \times .00237 \times 3376.4} \right)^{.35}$$

$$v = \left(\frac{39,600}{10,872} \right)^{.35} = (3.6423841)^{.35}$$

$$v = 88.4 \text{ fps} \quad \text{or} \quad 60.3 \text{ miles per hour}$$

Problem 2: What speed can a 14,000 cubic foot airship, with a pilot pedaling at .3 HP, expect to achieve? k = 242.41 (as proportionally scaled)

Setup:
$$v = \left(\frac{0.3 \times 550 \times .60 \times .40}{.0136 \times .00237 \times 242.41} \right)^{.35}$$

$$v = \left(\frac{39.6}{.00781} \right)^{.35} = (4937.8)^{.35} = 17.0 \text{ fps or } \underline{11.6 \text{ mph}}$$

(speed seems to be realistic)

Summary: From a study of the input numbers it is apparent that the speed of an airship is proportional to the cube root of the horsepower. If the pilot averages 0.3 Hp. over a course, he can only hope to change a few variables to increase his speed:
1. Lower his Coefficient of Drag *(Cd)* with a higher Fineness Ratio;
2. Streamlining the gondola to achieve a lower wind *(F)* resistance;
3. Increasing his power output (up to 0.7 *Hp*. is achievable in a sprint);
4. Increasing the propeller *(E)* efficiency (88% is achievable with technology).

Analysis: Based on actual tests, the formula (and result) seem *more* appropriate to micro-volumes and/or fractional horsepower inputs.

*NACA Report No.195; Airship Aerodynamics p25.

Cycle POWER Chart

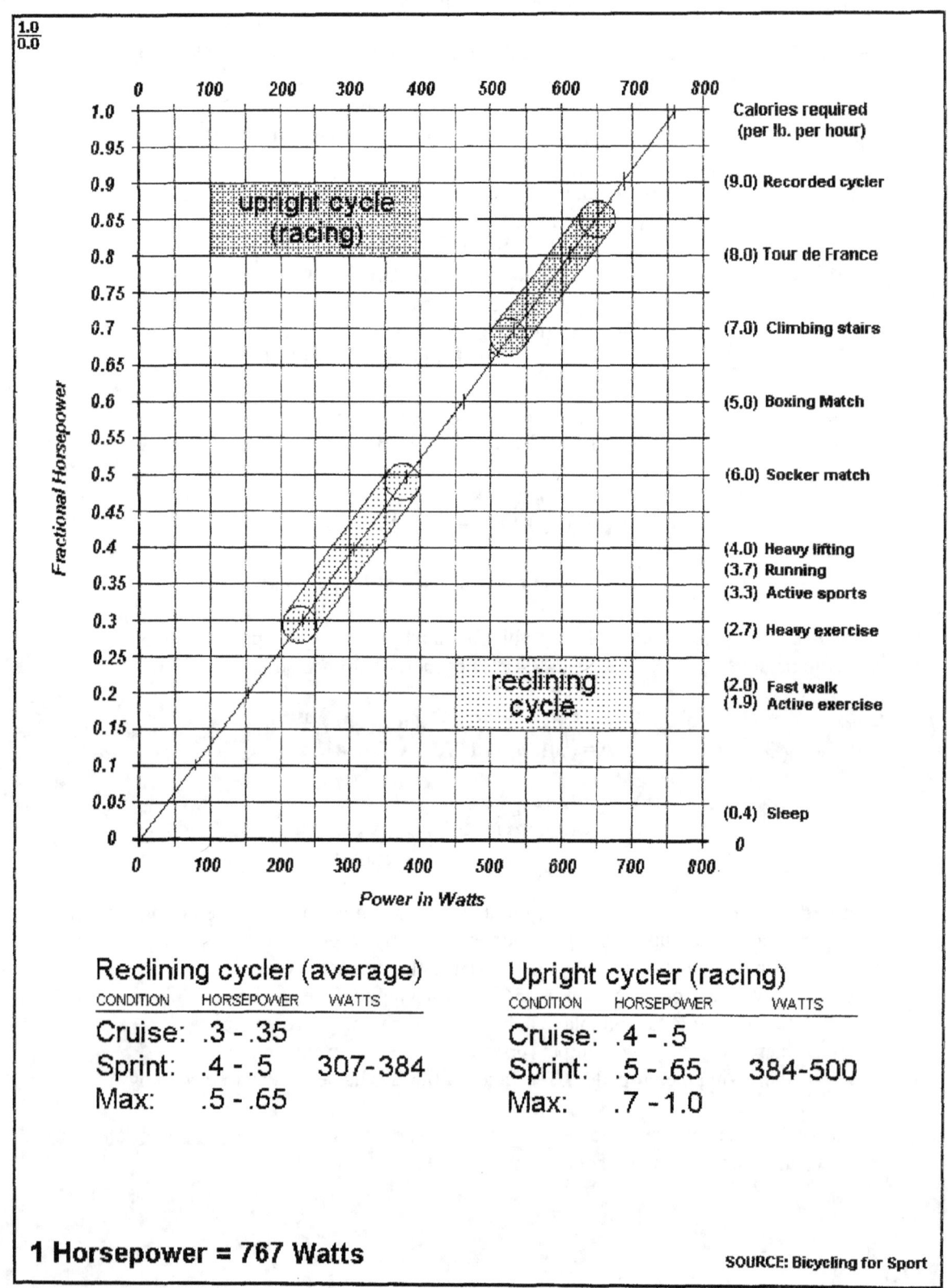

Reclining cycler (average)

CONDITION	HORSEPOWER	WATTS
Cruise:	.3 - .35	
Sprint:	.4 - .5	307-384
Max:	.5 - .65	

Upright cycler (racing)

CONDITION	HORSEPOWER	WATTS
Cruise:	.4 - .5	
Sprint:	.5 - .65	384-500
Max:	.7 - 1.0	

1 Horsepower = 767 Watts

SOURCE: Bicycling for Sport

BODY POSITIONS

1.2
0.6

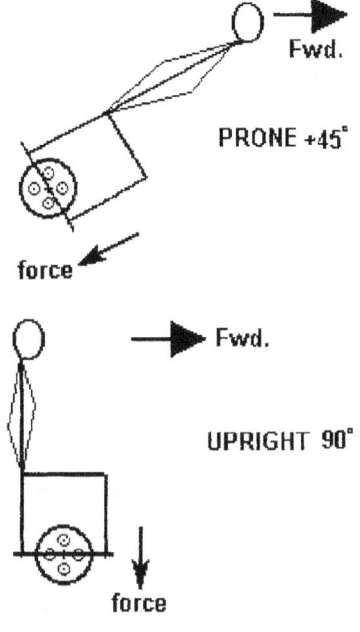

PRONE +45°

UPRIGHT 90°

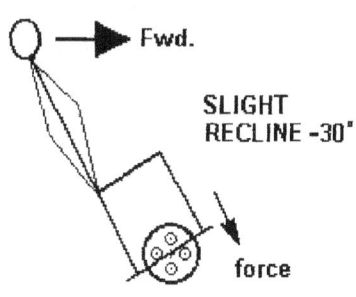

SLIGHT RECLINE -30°

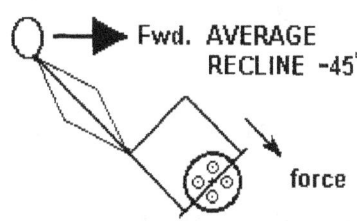

AVERAGE RECLINE -45°

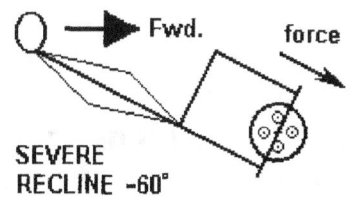

SEVERE RECLINE -60°

The best position should be the one making the best use of leg force and providing the most pilot comfort. The effects of streamlining on the reduction of body drag is negligible compared to the frontal area of the envelope.

PRONE POSITION: PRO: Good visibility; Has the best possibilities for aerodynamic streamlining.
CON: Uncomfortable chest and neck position. Reaching for controls becomes very limited. Very hard to mount & dismount.

UPRIGHT, PRO: Best power position; leg force uses full body weight; Not tiring; Controls easy to reach.
CON: Not relaxing over long duration's.

SLIGHT RECLINE: PRO: Downward force is still high in efficiency; Less fatiguing over long duration's.
CON: Bladder control becomes a problem.

AVERAGE RECLINE: PRO: More relaxing.
CON: Downward force decreases rapidly; but fatigue also decreases greatly. Bladder pressure becomes impossible to relieve.

SEVERE RECLINE: Very relaxing.
CON: Downward force inefficient as the force pushes the pilot weight away from the pedals. Harder to mount & dismount. Bladder control becomes a serious limitation.

GENERAL: Reclining & Prone positions require:
1. more attention to placement of controls;
2. adjustable seat position for multi-pilot comfort;
   (pilot weight, length, & fitness require dif. positions).
Higher rideing positions produce more instability
   in the gondola suspension.

COMPROMISES:
1. For speed, leg force is of primary concern.
2. For endurance, comfort is of primary concern.
3. For distance, visibility is of primary concern.

Non Rocket Science GONDOLAS

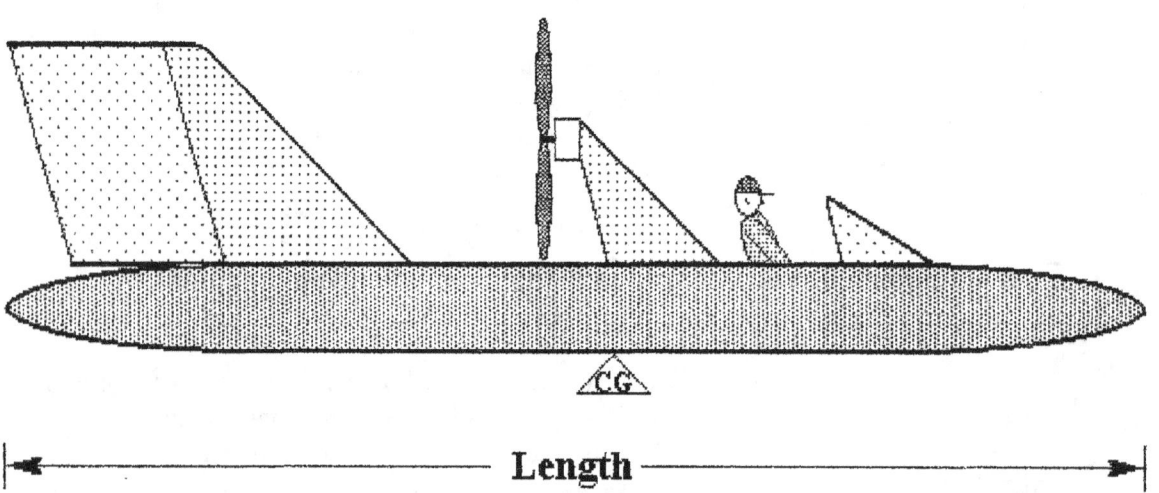

Targets
Simple, clean & functional controls; every part easily accessable
Target Weight less than: 120 pounds (54.5 Kilos)

Limitations
Safe for sustained human flight; able to support 3 times pilot wt.
Capable of quick disconnect from envelope
Capable of standing alone; easily transportable in sections
Reliable mechanically

Compromises
Preferrably a welded tube, or composit structure
Indoor or Outdoor use & storage (Ground Handling attachments)
Easily serviceable with commonly available parts

Working with Aircraft Tubing

Aircraft Tubing. There are three kinds of tubing in general aircraft use. *Steel tubing* in aircraft is almost invariably made of chrome-molybdenum (crom-oly) steel. The shape of the tubing may be either round, streamlined, square, or elliptical, though the elliptical shape is not common. *Aluminum-alloy tubing* is usually made from various alloys and is also available in standard shapes just mentioned, the most common being round and rectangle. *Stainless tubing,* alloyed with Chrome and nickel (18-8) is ordinarily supplied in the round shape only and is used almost exclusively for fuel and oil lines.

The size of tubing is designated by the outside diameter (OD) and the wall (W) thickness, the latter being designated by the Birmingham Wire Gage, or B.W.G. This rule applies in the case of round, elliptical, or streamlined tubing. In other words, since the streamlined tube is made from a round tube, the size of the round tube is specified. The dimensions of square tubing are usually expressed in terms of the width of one side.

Steel Tubing.—Steel tubing is used for the construction of fuselage frames, controls, bracing, and for any part where rigidity and strength are desired. It may be readily welded, which means that fittings can be attached without trouble. It may be flattened at the ends for various types of attachments; it may be bent and it may be given various shapes by the application of heat, which does not destroy its properties. Tubing made of 4130 steel has an ultimate tensile strength of about 100,000 pounds per square inch without heat treatment and up to 200,000 pounds per square inch after it has been properly heat treated. Any welding that is done must precede the heat treatment if this high strength is desired. The properties mentioned make 4130 steel tubing the ideal material for most structural parts.

Fitting Joints in Steel-Tube Structures.—In building a fuselage or other structure from steel tubing, one of the most difficult parts of the work is the matter of securing proper fits at the joints. A typical joint without reinforcements is shown in Fig. 9 (a). In this case three tubes are joined to a central tube after having been cut to the correct shape to make the joint; then the intersections between the intersections of the joint are welded as illustrated. It is important that the several tubes be fitted accurately, so that their center lines meet in a common point a, in order that the stresses shall be transmitted properly from one member to the other, and also to facilitate welding.

A cluster joint is sometimes reinforced, as illustrated in view (b). The reinforcement consists of a rectangular plate, or gusset, a that is inserted in a slot cut through the tube **b** and the branch tubes **c** and **d** on their center lines. All three tubes are then welded to the reinforcement, as shown at e. In addition, welds **f** are made between the tubes.

The hardest problem in making a good fit between two pieces of tubing is the scribing of the proper curve on the tubing which is to be cut. This problem may be greatly simplified by using the method shown in Fig. 10. The diagonal member a is blocked up at the proper angle, with one end in contact with the tube b to which it is to be fitted. This end may be held fixed by a light tack weld. On the upper side of the diagonal member a small pointed straightedge **c** is laid with the pointed end just touching the horizontal tube **b**. A mark is made on this straightedge and also on the diagonal member, as at **d**. The straightedge is then moved around the diagonal member, being kept parallel to the center line of the diagonal member at all times. At a number of points a mark is made on the diagonal in line with the mark **d**. After a series of such marks has been made completely around the diagonal member, the marks are connected by a curved line, as indicated at **e**. If the diagonal member is cut on this line, it will fit the other piece of tubing accurately. If the other end of the diagonal is also to be welded into the structure, an allowance of from 1/32nd inch to 1/16th inch, depending on the size of the tubing, is customarily made to compensate for expansion that may take place when heat is applied.

Aircraft Metalwork, Part 1 p26-32

MUSCLE POWERED BLIMPS

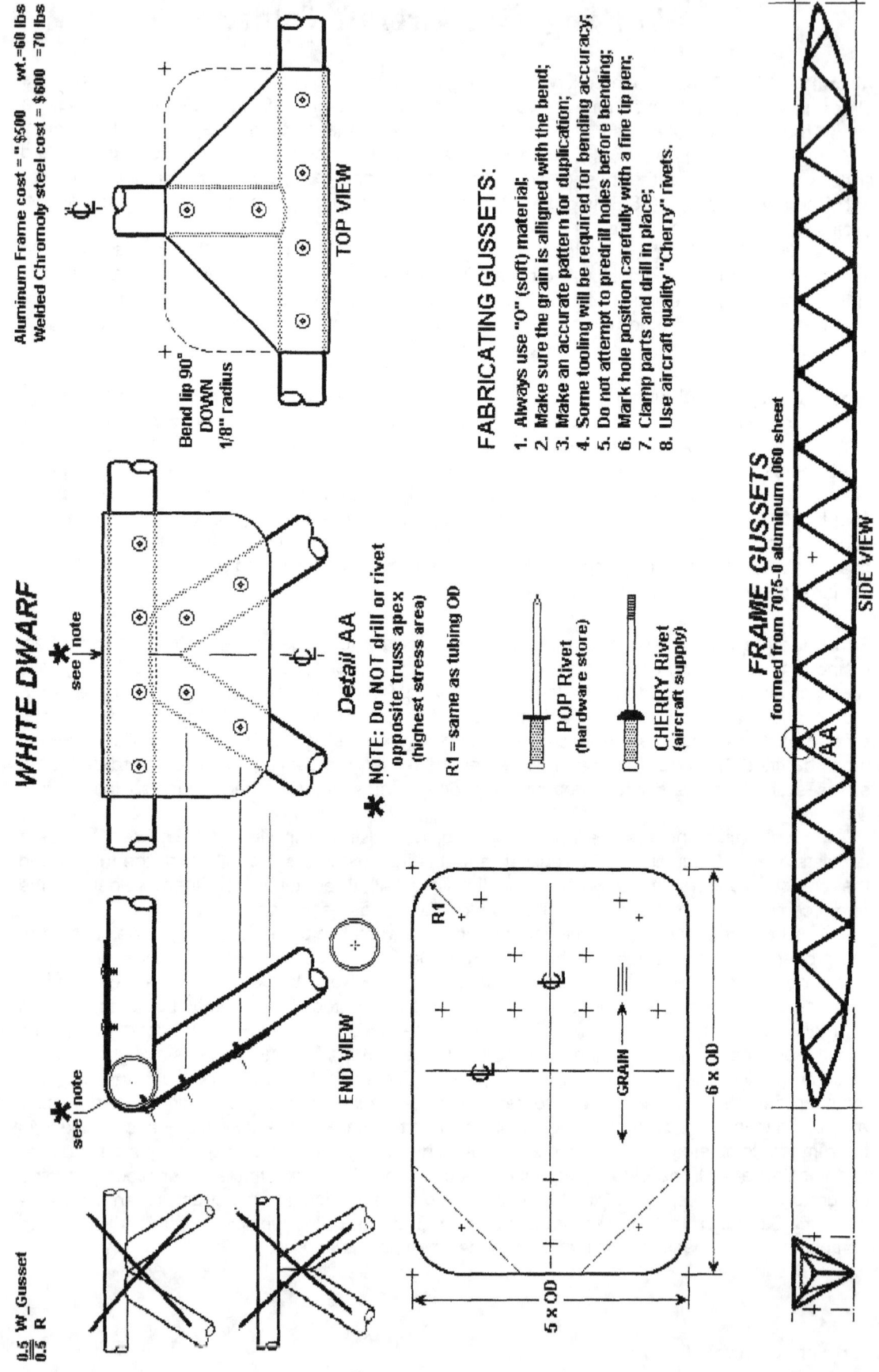

56

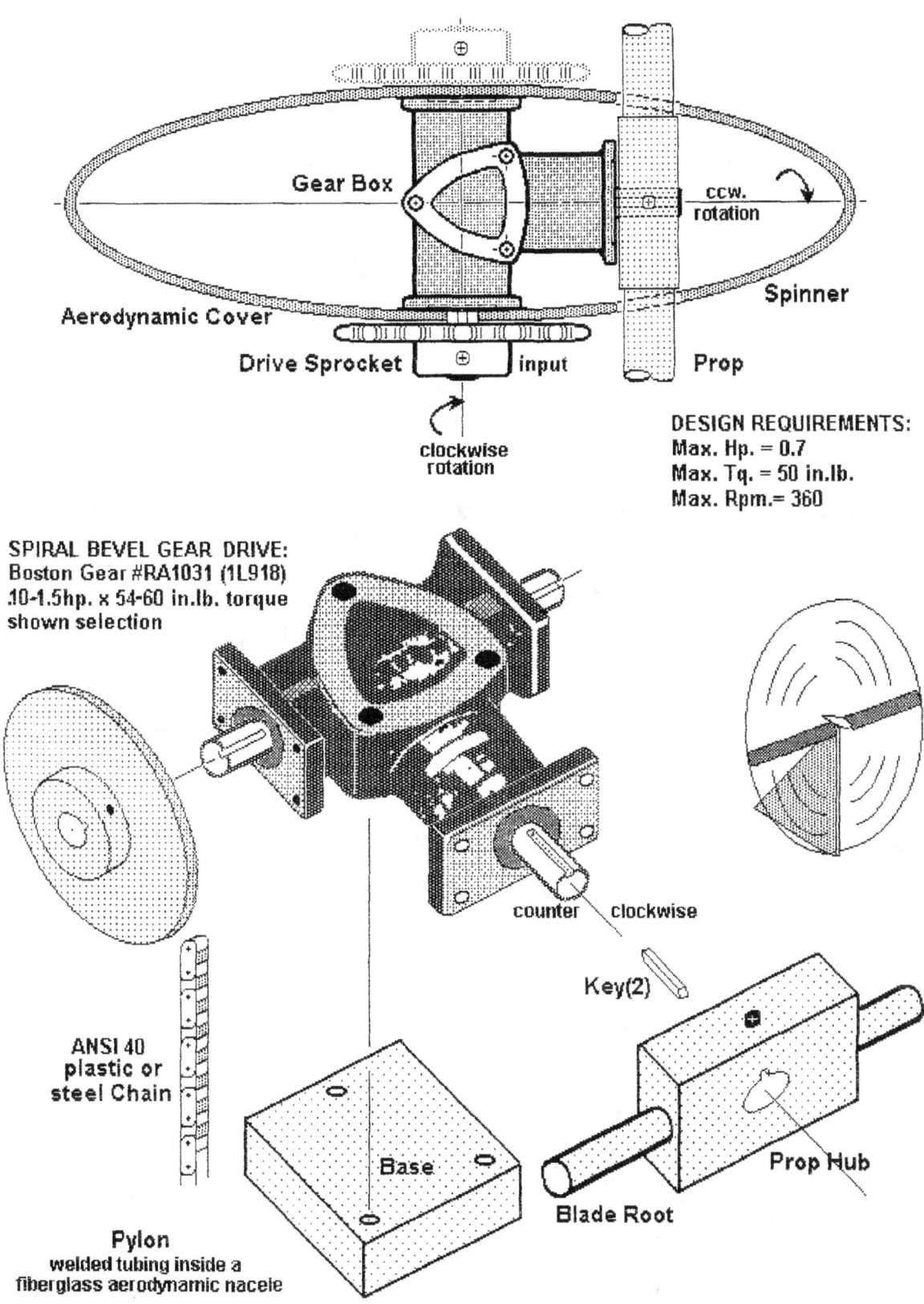

MUSCLE POWERED BLIMPS

BOSTON SPIRAL BEVEL GEAR DRIVES

2-Way — 1L910-1L915 series
3-Way — 1L916-1L921 series

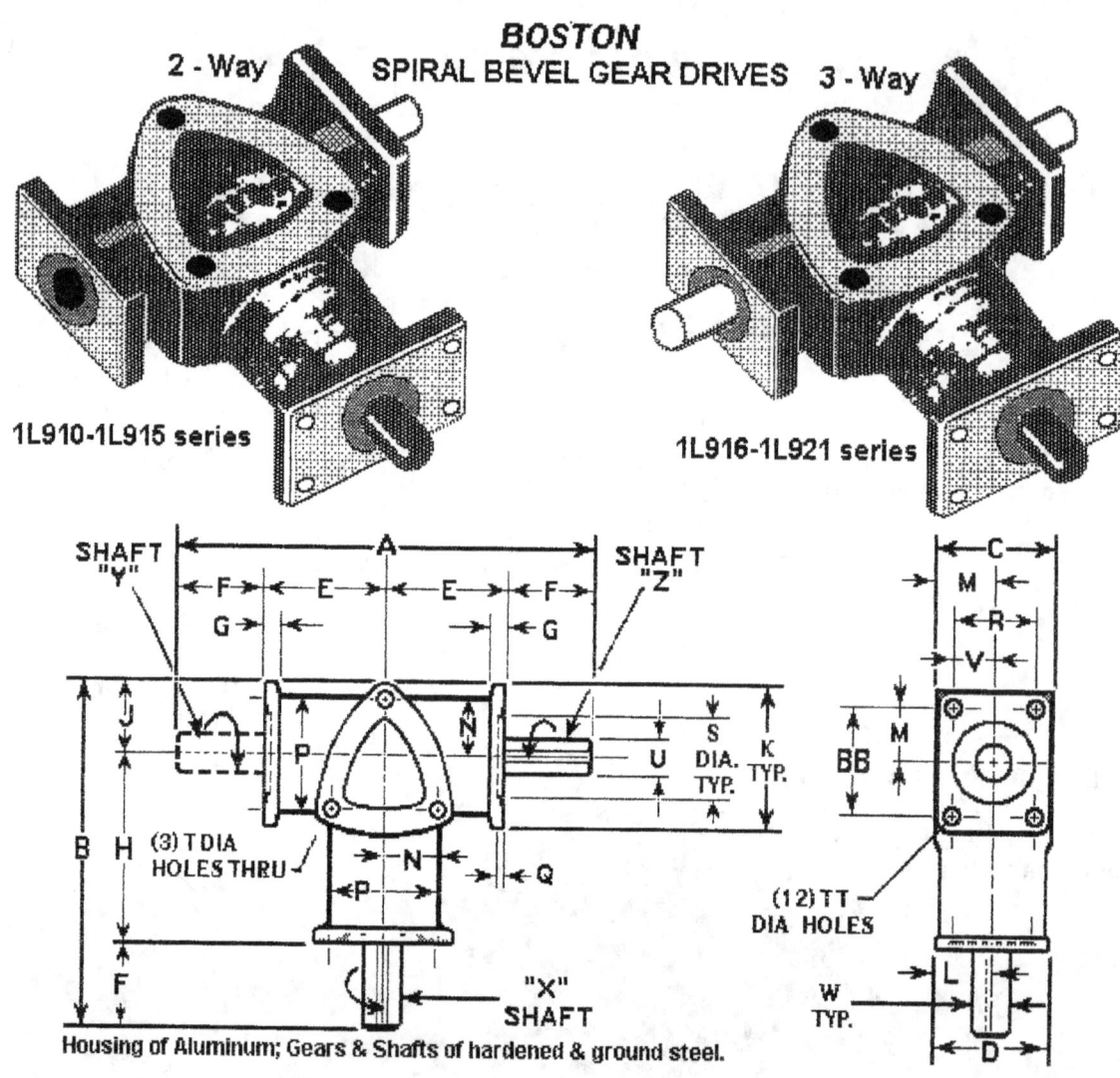

Housing of Aluminum; Gears & Shafts of hardened & ground steel.

DIMENSIONAL DRAWING FOR SPIRAL BEVEL GEAR DRIVES

Model No.	A	B	C	D	E	F	G	H	J	K	L	M
RA6	3 15/16	3 21/32	1 1/4	1 7/32	1 3/8	1 9/32	3/16	2 5/32	2 9/32	1 9/16	5/8	5/8
RA10	7 1/4	6 1/32	2	1 15/16	2 1/8	1 1/2	1/4	3 1/4	1 9/32	2 7/16	15/16	1
RA15	10	8 7/8	3	2 15/16	3	2	5/16	5	1 7/8	3 1/4	1 15/16	2

No.	N	P	Q	R	S	Holes F	Holes G	U	V	Key K	AA	BB
RA6	21/32	1 5/16	3/32	7/8	7/8	3/16	5/32	3/8	7/16	FL AT	0.6	1.2
RA10	15/16	1 7/8	3/32	1 3/8	1 3/8	17/64	17/64	5/8	11/16	3/16	.94	1.9
RA15	1 1/2	3	1/8	2 1/4	2 1/8	5/16	.5/16	3/4	1 1/8	3/16	1.5	3.0

SELECTION CHART -- RATINGS FOR 1.0 SERVICE FACTOR

Boston Series		RA621 & RA631			RA1021 & RA1031			RA1521 & RA1531		
Input RPM	Output RPM	Output HP	Output Torque	Stock No.	HP	Torque	Stock No.	HP	Torque	Stock No.
1750	1750	0.33	12.0		1.5	54.0		3.5	125.0	
1150	1150	0.27	15.0	1L910	1.0	56.0	1L912	2.5	135.0	1L914
690	690	0.18	16.0	1L916	0.62	57.0	1L918	1.6	142.0	1L920
100	100	0.03	19.0		0.10	60.0		0.28	174.0	

Ratio 1:1

ORDERING DATA

2-WAY SPIRAL BEVEL GEARS						3-WAY SPIRAL BEVEL GEARS				
Boston Model	Stock No.	List	Each	Shpg. Wt	Ratio	Model	Boston Stock No.	List	Each	Shpg. Wt
RA 621	1L910	$216.49	188.25	1.0	1:1	RA 631	1L916	216.49	188.25	1.3
RA 622	1L911	251.85	219 00	1.3	2:1	RA 632	1L917	251.85	219.00	1.3
RA1021	1L912	292.68	254.50	4.0	1:1	RA1031	1L918	292.68	254.50	3.5
RA1022	1L913	388.70	338.00	3.3	2:1	RA1032	1L919	388.70	338.00	3.4
RA1521	1L914	543.96	473.00	8.8	1:1	RA1531	1L920	543.95	473.00	9.0
RA1522	1L915	718.75	625.00	10	2:1	RA1532	1L921	718.76	625.00	8.9

MUSCLE POWERED BLIMPS

Power Transmission
ROLLER CHAIN SPROCKETS

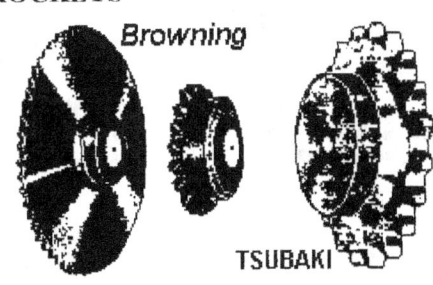

For use with Nos. 40 pitch (.50) roller chain.
Precision machined of high alloy carbon steel.
Sprockets with 30 teeth or less have
hardened teeth for increosod life.
Outside diameter (OD) range: 2.65" to 23.46".
Bore range 1/2" to 3 7/16" with required split
taper bushing (not included); order split taper
bushing separately on page 341.
Roller Chain available on page 317.

BUSHED Bore					ANSI SPROCKET No.40			(0.50 pitch)
OD	Pitch Dia.	No.of Teeth	Bush. Req'd.	Browning Model	Stock No.	List	Each	Shpg. Wt.
2.65"	2.405"	15	H	H40H15	6L527	44.76	17.98	0.6
2.80	2.663	16	H	H40H16	6L528	46.56	18.70	0.7
2.96	2.721	17	H	H40H17	6L529	60.16	20.17	0.7
3.14	2.879	18	H	H40H18	1L194	53.16	21.38	0.7
3.30	3.038	19	H	H40H19	1L195	66.36	22.64	0.8
3.45	3.196	90	H.	H40H20	1L196	69.12	27.85	0.9
3.62	3.656	21	P1	H40P21	6L530	61.94	24.92	1.6
3.75	3.613	22	P1	H40P22	6L531	69.14	27.90	1.7
3.94	3.672	23	P1	H40P23	6L532	73.56	29.60	1.8
4.10	3.831	24	H	H40H24	1L197	88.72	35.80	1.2
4.26	3.989	25	P1	H40P25	6L533	81.56	32.90	1.9
4.U	4.148	26	P1	H40P26	6L534	85.16	34.35	2.0
4.58	4.307	27	P1	H40P27	6L535	86.98	35.00	2.1
4.74	4.466	28	P1	H40P28	6L536	89.66	36.05	2.2
4.90	4.626	29	P1	H40P29	6L537	91.40	36.85	2.3
5.06	4.783	30	H	H40H30	1L198	96.32	38.80	1.7
5.38	6.101	32	P1	40 P32	6L538	101.56	40.85	2 6
5.86	5.578	35	P1	40 P35	6L539	104.76	42.20	2.9
6.02	6.737	36	P1	40 P36	1L199	110.42	44.50	3.1
6.65	6.373	40	P1	40 P40	1L200	120.62	48.55	3.6
6.97	6.691	42	P1	40 P42	6L540	122.82	49.45	3.6
7.45	7.168	45	P1	40 P45	1L201	125.02	50.35	4.3
7.93	7.645	48	P1	40 P48	1L202	130.02	52.35	5.0
8.89	8.599	54	P1	40 P54	6L541	136.22	54.85	5.3
9.84	9.554	60	Q1	40Q60	1L203	114.02	4610	7 7
11.43	11.15	70	Q1	40Q70	1L204	151.10	60.75	9 4

FIXED Bore				ANSI SPROCKET No.40				(0.50 pitch)	
	Pitch	No.	UST	For bore size specify Stock #					Shpg.
OD	Dia.	Teeth	Model	1/2"	5/8"	3/4"	7/8"	Each	Wt.
1.674"	1.462"	9	40B 9F	6L835	6L836	—	—	7.37	0.2
1.839	1.618	10	40B10F	1L104	1L105	1L106	—	7.64	0.3
2.003	1.775	11	40B11F	6L837	6L838	6L839	6L840	7.98	0.4
2.166	1.932	12	40B12F	1L108	1L109	1L110	1L111	8.74	0.4
2.328	2.089	13	40B13F	6L841	1L113	1L114	1L115	9.15	0.5
2.490	2.247	14	40B14F	1L117	1L118	1L119	1L120	9.63	0.7
2.652	2.405	15	40B15F	6L842	1L122	1L123	1L124	10.39	0.7
2.814	2.563	16	40B16F	—	1L126	1L127	1L128	11.00	0.8
2.974	2.721	17	40B17F	—	1L130	1L131	1L132	11.63	0.9
3.136	2.879	18	40B18F	—	1L134	1L135	1L136	12.25	1.0
3.292	3.038	19	40B19F	—	6L847	1L138	6L848	13.63	1.0
3.457	3.196	20	40B20F	—	6L850	1L140	1L141	15.61	1.4
3.618	3.355	21	40B21F	—	6L852	6L853	6L854	16.99	1.8
3.778	3.513	22	40B22F	—	6L857	6L858	6L859	18.64	1.7
3.938	3.672	23	40B23F	—	6L862	6L863	6L864	20.64	1.7
4.098	3.831	24	40B24F	—	6L867	1L143	6L868	22.14	1.7
4.258	3.989	25	40B25F	—	6L870	6L871	6L872	24.08	1.9
4.418	4.148	26	40B26F	—	6L875	6L876	6L877	25.60	1.9
4.738	4.465	28	40B28f	—	6L880	6L881	6L882	29.35	2.1
5.057	4.783	30	40B30F	—	6L885	6L886	6L887	32.05	2.3
5.856	5.578	35	40B35F	—	—	—	—	32.05	2.7

Standard keyways: 1/2-9/16 = 1/8 x 1/16 x 10-24; 5/8-7/8 = 3/16 x 3/32 x 1/4-28 set screw.

W.W. GRANGER, nationwide; 1998 catalog, pgs. 310-341

DRIVE MANDRELS

Sprocket Not Included

- 3000 RPM Maximum
- Precision setscrew-type ball bearing pillow blocks are pre-lubricated and sealed for maintenance-free operation
- Includes hex nut and two washers (1½" OD) for each end
- All components are plated to resist rust
- Suitable for jackshaft applications

Drive Shaft (In.)	Threaded Ends RH	LH	OD Pulley	Stock No.	List	Each	Shpg. Wt.
5/8 x 12	1/2-20	1/2-20	2"	6L098	$50.40	**$35.55**	3.3"
3/4 x 12	5/8-11	5/8-11	2"	6L099	50.40	**35.60**	3.6

ELASTOMERIC TENSIONERS

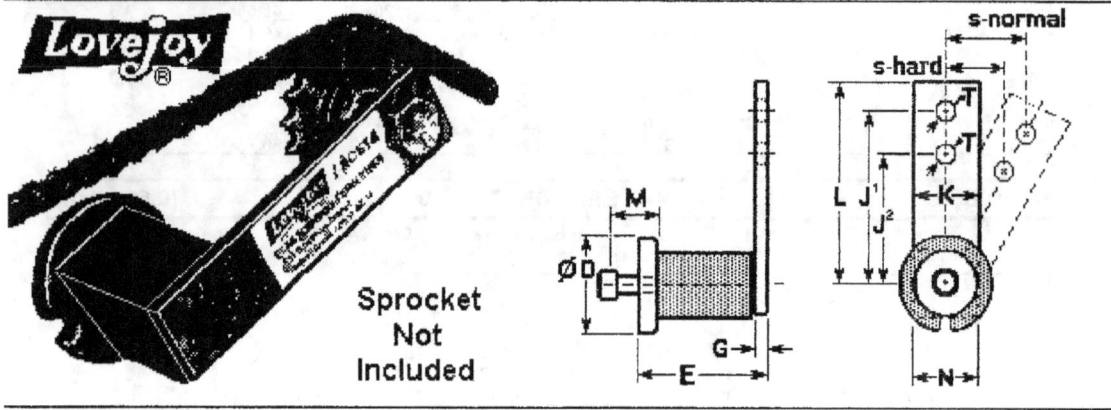

Sprocket Not Included

ORDERING DATA

ANSI Chain Size	Belt Range	Rosta Model	Stock No.	List	Each	Shpg. Wt.
25	A	SE11	3KZ08	$61.60	**$32.50**	0.7
35	A, B, 3L.	SE15	1L833	68.00	**36.20**	1.2
40, & 41	B, C, 4L, 5L.	SE18	1L834	82.00	**43.55**	1.8

Stock number	DIMENSIONS								Bolts	
	D	E	G	J1	J2	K	L	N	T	Metric	Std. Size
3KZ08	1.38	2.01	0.20	3.15	2.36	0.79	3.54	0.87	0.39	M6 x 20	3/8-16 x 2"
1L833	1.77	2.52	0.20	3.94	3.15	0.98	4.43	1.18	0.53	M8 x 20	1/2-13 x 2"
1L834	2.28	3.07	0.24	3.94	3.15	1.18	4.53	1.38	0.53	M10 x 30	1/2-13 x 2.5

POWER TRANSMISSION: ROLLER CHAIN

CORROSION-RESISTANT RIVETED ROLLER CHAIN AND INDIVIDUAL LINKS

RIVETED ROLLER CHAIN
SPECIFICATIONS AND ORDERING DATA

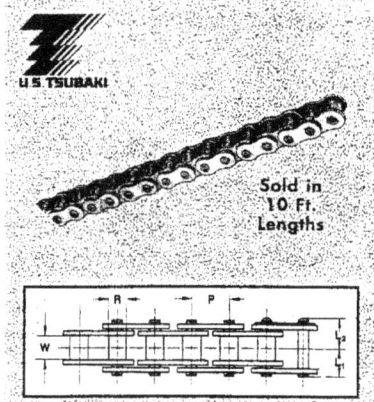

Sold in 10 Ft. Lengths

ANSI Size	Pitch P	Width W	Dia. R	Connecting End to Center Line L_1	Rivet End to Center Line L_2	Overall Width L_1 & L_2	Maximum Allowable Load (Lbs.)	Stock No.	List	Each 10 Ft. Length	Shpg. Wt.
40PC	1/2	5/16	0.312"	0.325"	0.392"	0.717"	100	6L079	$178.00	$153.50	2.6
50PC	5/8	3/8	0.400	0.406	0.472	0.878	154	6L080	223.00	192.75	4.0
60PC	3/4	1/2	0.469	0.506	0.581	1.087	198	6L081	311.00	268.75	6.0
40NP	1/2	5/16	0.312	0.325	0.392	0.717	660	6L073	42.00	36.25	4.5
50NP	5/8	3/8	0.400	0.406	0.472	0.878	1140	6L074	60.00	51.40	7.0
60NP	3/4	1/2	0.469	0.506	0.581	1.087	1630	6L075	79.00	67.85	11.0
40SS	1/2	5/16	0.312	0.325	0.392	0.717	100	6L076	209.00	180.50	4.4
50SS	5/8	3/8	0.400	0.406	0.472	0.878	154	6L077	262.00	226.75	7.0
60SS	3/4	1/2	0.469	0.506	0.581	1.087	231	6L078	366.00	316.25	10.0
40AS	1/2	5/16	0.312	0.325	0.392	0.717	155	6L070	275.00	237.75	4.3
50AS	5/8	3/8	0.400	0.406	0.472	0.878	230	6L071	363.00	313.75	7.0
60AS	3/4	1/2	0.469	0.506	0.581	1.087	350	6L072	494.00	426.50	10.0

CORROSION RESISTANCE GUIDE

Substance	SS AS	NP	PC	Substance	SS AS	NP	PC
Acetone	•	=	•	Carbon Tetrachloride	#	=	#
Oil (Plant, Mineral)	•	•	•	Potassium Hydroxide (20%)	•	=	=
Alcohol	•	=	•	Sodium Hydroxide (20%)	•	=	=
Ammonia Water	•	=	•	Nitric Acid (5%)	=	=	=
Sodium Chloride	#	=	#	Vinegar	#	=	#
Hydrochloric Acid (2%)	=	=	=	Hypochlorite Soda	=	=	=
Sea Water	#	=	#	Soft Drinks	•	=	=
Hydrogen Peroxide	=	=	=	Soap and Water Solution	•	#	•
Caustic Soda (25%)	•	=	=	Paraffin	•	•	•
Gasoline	•	•	•	Fruit Juice	•	=	•
Formic Acid	=	=	=	Benzene	•	=	•
Formaldehyde	•	=	•	Water	•	#	•
Milk	•	•	•	Vegetable Juice	•	=	•
Lactic Acid	•	=	•	Iodine	=	=	=
Citric Acid	•	=	#	Sulfuric Acid	=	=	=
Chromic Acid (10%)	•	=	=	Phosphoric Acid (10%)	#	=	=
Acetic Acid (5%)	•	=	•				

• Highly corrosion resistant.   # Marginally corrosion resistant (depending on application conditions).
= Not corrosion resistant.

POLY CHAIN (PC)
Designed for conditions where lubrication is impractical or undesirable. Can withstand some corrosive conditions and is suited for environments requiring sanitary conditions. Inner links are polyacetal, outer links are 304 stainless steel.

NICKEL PLATED (NP)
Suitable for slightly corrosive environments and for outdoor conditions exposed to rain. Temperature range: 15°F to 140°F. Not recommended in applications exposed to food or high temperature variance.

304 STAINLESS STEEL (SS)
Ideal for fresh and saltwater exposures. Suitable for corrosive conditions involving food, chemicals, and pharmaceuticals. Temperature range: -40°F to 750°F.

600 STAINLESS STEEL (AS)
Combines corrosion resistance of 304 stainless steel with a 50% higher maximum allowable load made possible through heat treating of pins, bushings, and rollers. Temperature range: -40°F to 750°F.

For more detailed information on specific substance compatibility, see "Corrosion Resistance Guide" above right.

A WIDE SELECTION OF BELTS IS AVAILABLE, SEE PAGES 324 THRU 331.

INDIVIDUAL CHAIN LINKS
(SOLD IN PACKAGES OF 5)

 Connecting Link  Offset Link

ANSI Size	CONNECTING LINKS Stock No.	List	Per Pkg.	Shpg. Wt.	OFFSET LINKS Stock No.	List	Per Pkg.	Shpg. Wt.
40NP	6L091	$10.00	$8.07	0.1	6L082	$19.00	$16.30	0.1
50NP	6L092	12.00	10.25	0.3	6L083	25.00	21.05	0.2
60NP	6L093	20.00	16.34	0.3	6L084	30.00	25.30	0.3
40SS*	6L094	21.00	16.38	0.1	6L085	35.00	29.85	0.1
50SS*	6L095	23.00	19.15	0.2	6L086	41.00	35.35	0.2
60SS*	6L096	29.00	24.32	0.3	6L087	57.00	49.15	0.3
40AS	1L541	22.00	18.65	0.1	6L088	44.00	37.35	0.1
50AS	1L540	26.00	22.37	0.2	6L089	52.00	44.30	0.2
60AS	1L539	36.00	30.70	0.3	6L090	71.00	61.40	0.3

(*) Also for use on Nos. 6L079 through 6L081.

 Chain Puller   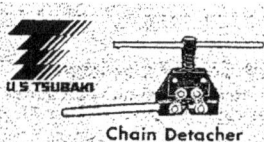 Chain Detacher

ROLLER CHAIN TOOLS

Description	ANSI Size	Mfr.	Mfr's. Model	Stock No.	List	Each	Shpg. Wt.
Chain Puller	35-60	Browning	35	5A555	$25.18	$17.98	0.4
Chain Puller	80-240	Browning	80	5A556	47.83	34.30	2.2
Chain Detacher	35-50	UST	D-35	1A911	31.00	26.35	0.7
Chain Detacher	60-100	UST	D-60	1A912	58.00	49.65	1.6

Non Rocket Science
PROPELLERS

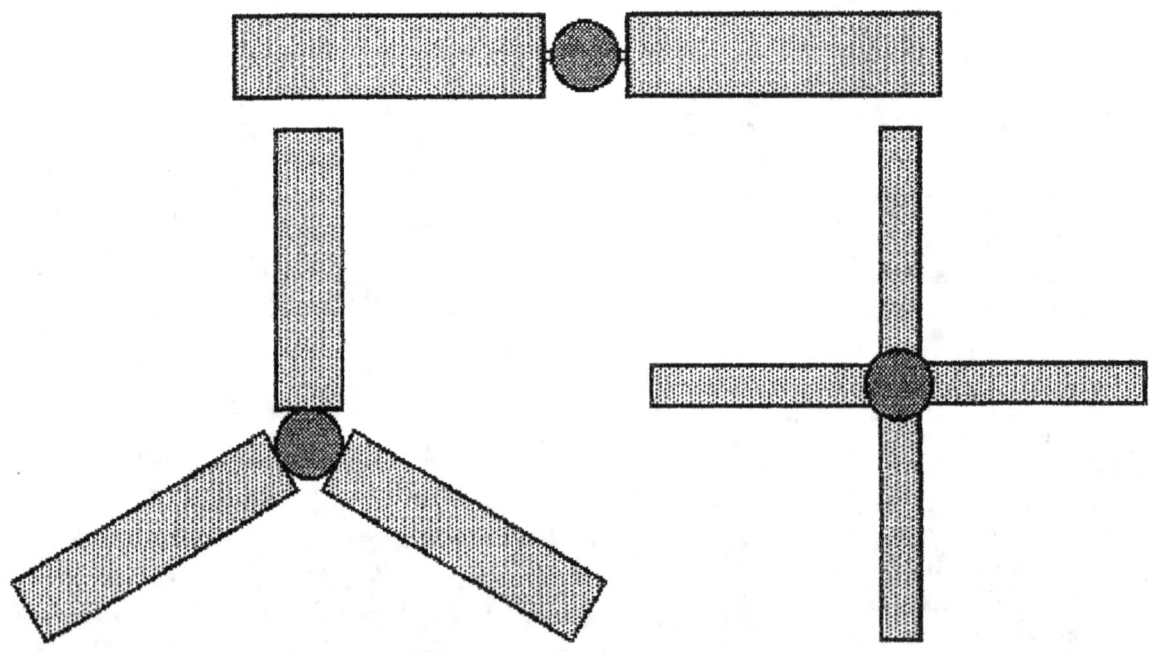

Targets
Blimp Target Speed: 15 MPH (22 Feet per Second)
Prop Target Tip Speed: Mach .5 to .67 (5600 to 7500 FpS)

Limitations
Envelope Frontal Area: 282 square feet (Class BA1 at 4:1)
Envelope shape Drag Coefficient: .0136
Prop Diameter Range: 5.5' to 6.5' (clear. betw. gondola & envl.)
Shaft Horsepower: .25 to .35 cruise (4.0 to .5 max. sprint)

Compromises
Shaft RPM: 350 to 360 (chain & gearbox limits)
Number of Prop Blades: 2 to 4 (blade resistance & lift distribution)
Blade Chord length: 4 to 10 inches (ease of fabrication)

PROPELLERS

Well, we've explained some of the rocket science, but arn't quite ready to start cutting the wood and foam yet. By now I hope you understand that we are virtually off the mathamatical scale, on the very low end of propulsion technology. It actually takes very little thrust to move a small blimp around inside a hangar. But where the men are separated from the boys is when the hangar doors are opened. The great leap outside into the sky is suddenly frought with a new set of challenges; Some, of which can easily be life threatening. Lets look at the problem a little more closely.

Pilot Limitations:
 What is his physical condition? What kind of pedal power and endurance can he sustain? And for what period of time? What is he capable of in a 2-minute sprint? I bring this up as a fair warning; ***The Wind is Not your Friend***.

Weather Limitations:
 What are his personal limits on what wind conditions you will fly in? Or even take the airship out of the hangar in? Would you allow someone to talk you into going out in a 15 knot wind when you know you can only pedal 10?

Being Realistic: These numbers may vary slightly with individuals and design, but lets consider the following: (Challenge the numbers with your own Ergometer, if you like)
1. A bicycle rider in good shape is capable of the following:
   a. A maximum horsepower of about 0.6 for 15 seconds;
   b. A sprint of about 0.4 for 2 minutes;
   c. A continuous cruise of .25 for 10 minutes.
   d. 10 seconds of coast is required in each 60.
2. Since airship stability is dependant on how close the centerline of thrust is to the longitudinal axis, the prop diameter will probably be 5 to 6 feet.
3. Since bicycle chain and sprockets will most likely be used, the input RPM will be in the range of 60-80, and the output 300-400 in cruise.
4. Since a 5:1 gear ratio is probably about optimum, a blade chord (width on a 5.5' prop length) will be between 8 and 10 inches.
5. With so many variables, be prepared to make and test several different props.
   a. keep them simple with a rectangle shape (no taper, constant chord);
   b. make your first one in one piece, a Clark-Y, with a 7.5 pitch angle;
   c. Don't try exotic designs without knowing the characteristics;
6. Knowing you are going to make more than one:
   a. start with a written plan, and make layout templates;
   b. make blades in a mold if possible;
   c. have a way check that your dimensions are consistant;
   d. keep the weights and balance of blades as exact as possible.

PROPELLERS

As I see it, there are two ways to approach this. Either by "Rocket Science" or "Cut and Paste". If the former is over your head, then by all means, turn to the latter pages and start cutting wood and foam.

Rocket Science is a good way to approach the subject, just to review some of the terms we will be working with. Don't let the numbers boggle you to a point you loose sight of the target; that of getting into the air and making it go. Since I don't know the motivation for your project, I can only assume your aircraft is for personal pleasure. If the ultimate target happens to be a commercial endeavor, or for some exotic mission (like for setting a new World record), then you had better get some serious engineering help. So for now, lets carry on.

A **propeller** is a rotating airfoil used to create thrust. They have been around a long time, and have developed to very efficient machines over the last 100 years. But not without an understanding of what makes it efficient. A rotating brick (two are better) can produce thrust, but an aerodynamic shape will work better.

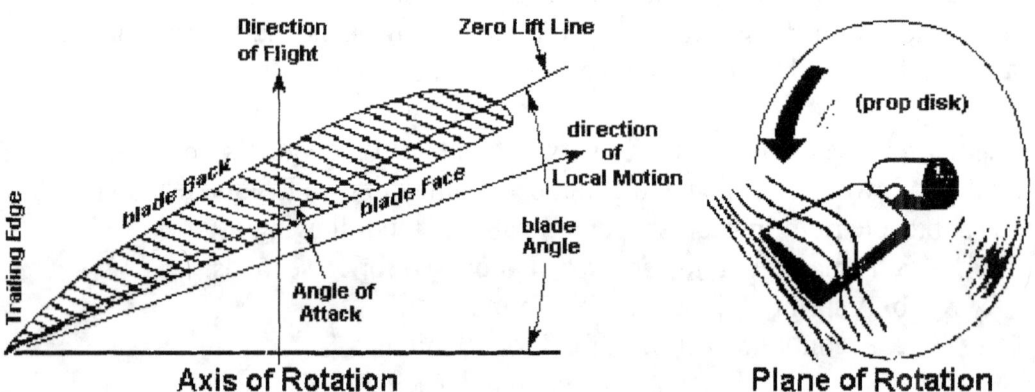

The aerodynamic *shape* obviously gives a better reaction to using the air. As the *blade angle* is increased, it is apparent that a greater bite is made on the medium it is passing through. When we refer to a specific shape, we might say that it is a "Clark-Y" or a "NACA-4412" (or some other). What that is telling you, is that the shape has been tested, and has specific known results of performance. A four digit NACA number is telling you the % of maximum thickness, as well as the location of that max.thickness (and any camber).

Lest we miss an option available later on, I will mention that there are three Types of propellers: ***Fixed Pitch***, made of one piece, are most commonly found on light aircraft. ***Ground Adjustable Pitch***, with blades that can be set for specific conditions. And *Controllable Pitch*, adjustable in flight, used on complex aircraft. This chapter on props will only discuss 2-bladed, Fixed pitch. However, an adjustable pitch prop can positive advantages on an ultra-light aircraft. A Multi-blade prop should also not be overlooked. It can also be of great benifit.

Prop.Theory, p5-6.

PROPELLERS

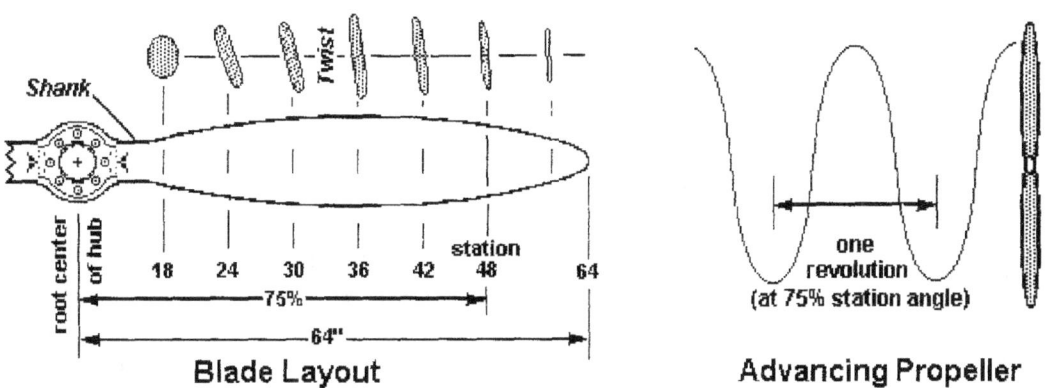

Blade Layout   **Advancing Propeller**

NOTE: *Blade Twist* is shown for illustrative purposes. It may not be desireable on small props. But it does result in better efficiency.

Black Art; Design inputs and formulas are available to optimize the performance of a propeller design. It is called "black" art, as the design calculations seldom meet actual performance. The numbers on higher speed aircraft become fuzzy when the blade tip speed approaches Mach-0.7. Low speed aircraft numbers get fuzzy from a myriad of compromises. And very low speed aircraft (less than 15 mph) usually suffer from prop lenght restraints that limit the best use of (fractional) horsepower available.

The inputs that you must know are:
    Horsepower (Hp) of the power source;
    Speed (RPM) of the prop shaft;
    Radius (d) of the prop disk;
    Advance Ratio or Pitch of the prop. (AR=Pitch/Diameter)
From these, you will need to know:
    Tip Speed (V_t) of the prop in ft./sec. (.52 x Diameter x RPM);
        Not to exceed 638 for an all wood prop.
    Velocity (True Velocity, T_v) at 75% of the blade Radius
    Sine of the Blade Angle.

From these, you can find the approximate efficiency (E) of the prop. E= Thrust in lbs, times the aircraft speed in feet per second, divided by Power in foot pounds per second.

When the aircraft is complete, and under max power, the other loses are caused by parasite drag of the blade, friction in the drive system, energy lost in the slipstream(*), and everything disrupting the airflow through the prop. A "spinner" on the prop hub tends to smooth the flow around any cowling.

NOTE: 1 Horsepower = 550 ft.lb./sec = 767 Watts.
* see the "ELICA-Larrabee Lecture" under "Rocket Science"

Prop.Theory, p5-6.   R2

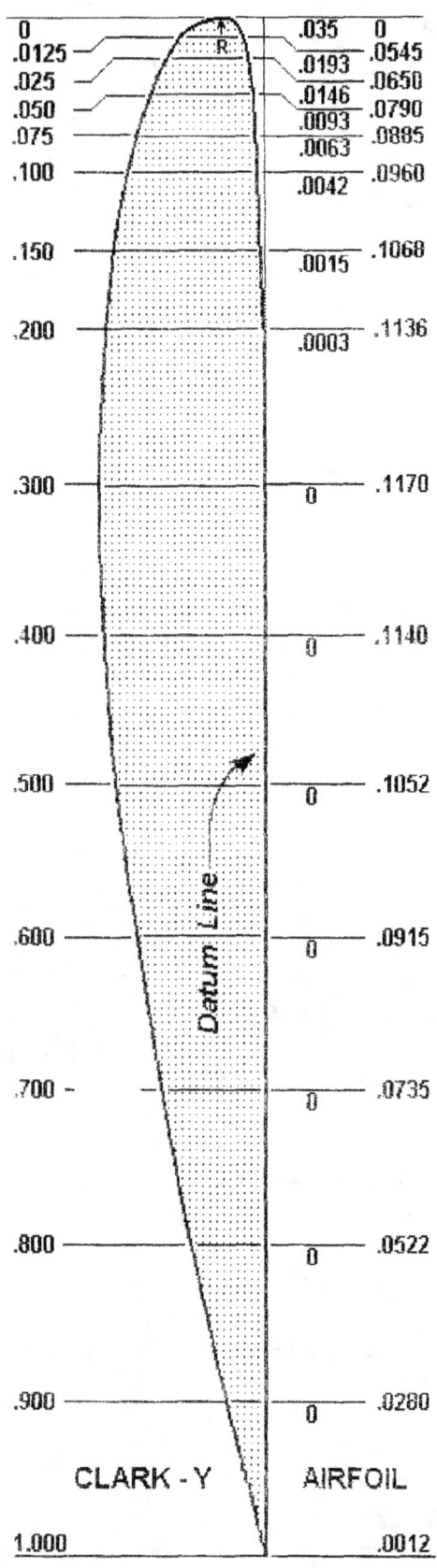

PROPELLERS

AIRFOIL PROFILE LAYOUT

The upper and lower camber or curvature of an airfoil section is laid out from abscissae and ordinates supplied by the wind tunnel or aerodynamic laboratory which tests the airfoil. They are always given in decimal parts of the chord. Thus, a rib of any size may be laid out in proper proportions. The abscissae are measured from the nose or leading edge and the ordinates from a datum line, which is usually a line touching the bottom curve at two or more points but in the case of double-cambered sections may run from the nose to the tail or trailing edge.

To obtain the proper dimensions in inches for laying out the section to scale, the section to scale, abscissae and ordinates must, of course, be multiplied by the chord length in inches.

The airfoil which will be used below is the Clark Y, an excellent section for all around use. The chord length chosen is ten (10") inches.

The lowest row of figures indicates the distance from the nose in decimal parts of the chord, the next row is the height of the bottom curve from the datum line and the top row the height of the top curve. It will be noted that from .3 on back the bottom ordinate is zero, indicating that the bottom of the section is flat. The next step is to lay out a table of ordinates for convenience. The procedure is clearly indicated on the following page.

Secure a piece of bend resistant .080 aluminum sheet, spray it with Prussian blue, and using a straight edge, mark the datum line with a scribe. With a vernier calipers, lay out the abscissae (second column) along the line. If such a scale is not available, a table of decimal equivalents must be and the dimensions converted to thousandths.

EFFECT OF THE PROPELLER ON PERFORMANCE

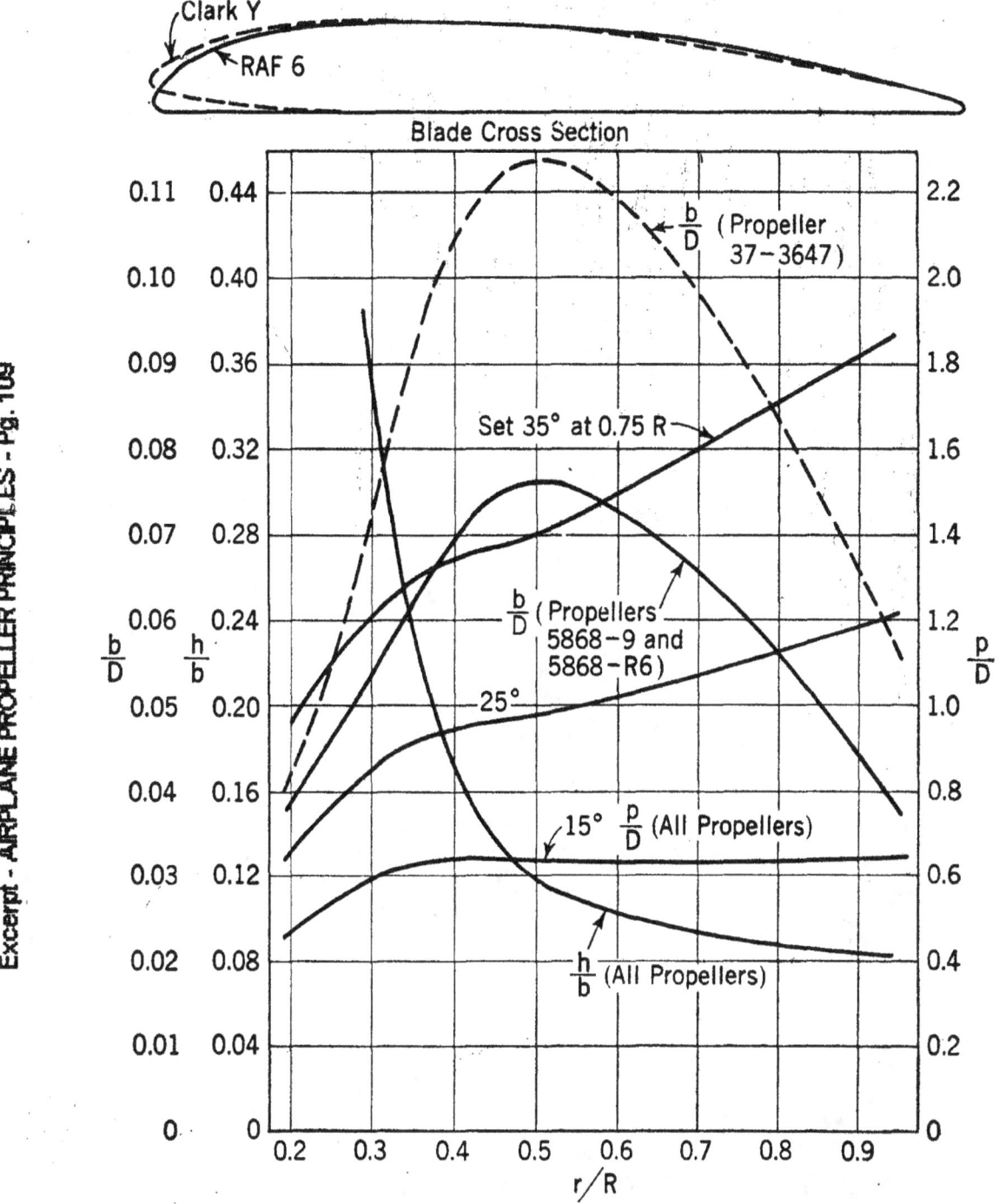

FIG. 5·4. Blade form curves. (*NACA Tech. Rept. 640.*)

MUSCLE POWERED BLIMPS

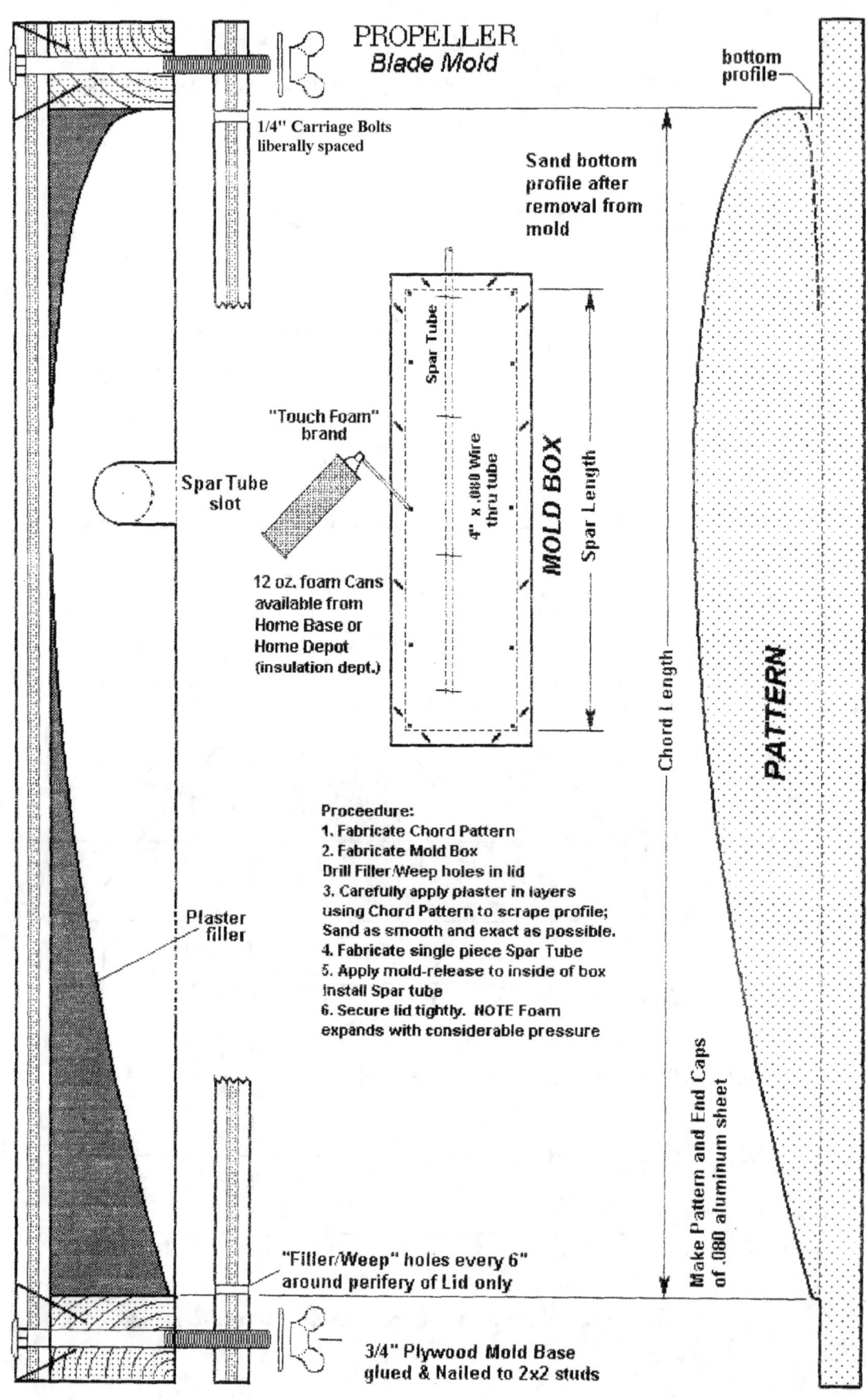

PROPELLERS

PROFILE COORDINATES

Distance from Nose	Times 10"	Bottom Camber	Times 10"	Top Camber	Times 10"
.0	0	.035	.35	.035	.350
.0125	0.125 "	.0193	.193	.0545	.545
.025	.25	.0146	.146	.0650	.650
.050	.50	.0093	.093	.0790	.790
.075	.75	.0063	.063	.0885	.885
.100	1.00	.0042	.042	.0960	.960
.150	1.50	.0015	.015	.1068	1.068
.200	2.00	.0003	.003	.1136	1.136
.300	3.00	.000	0	.1170	1.170
.400	4.00	.000	0	.1140	1.140
.500	5.00	.000	0	.1052	1.052
.600	6.00	.000	0	.0915	.915
.700	7.00	.000	0	.0735	.735
.800	8.00	.000	0	.0522	.522
.900	9.00	.000	0	.0280	.280
1.00	10.0	.000	0	.0012	.012
R	**Radius**				
.015	.150				

Square up a line accurately from each point and set of the ordinates (columns 4 and 6). It is desirable to use a double needle compass to mark the leading edge circle, and a steel scribe to connect the curve points.

Bend a smooth, flexible piece of wood or plastic so that it lies with one edge directly through the points (or at least 3 points in a row). It may be held in position by tape, weights, or an assistant. With a sharp pointed scribe, draw a line through the points of top or bottom curve, then through the other set of points. With a Compass set to the proper radius (.150" in this case) draw the nose radius just tangent to the curve.

It will be noted that the instructions above call for the use of a strip of flexible material (called a "spline" or "batten") in laying out the curve. It is possible to get long sweeping curves made plastic or metal called "French, or ship curves" which will usually fit the lines to be drawn and there is always a temptation, if they are available, to use them instead of a spline. This should never be done, however, as it is practically impossible to get an absolutely "fair" or smooth curve, with no irregularities, unless a spline is used first. If it is desired to make the line heavier, the ship curves may be used afterwards.

Bear in mind that accuracy in a wing section is of utmost importance, and it is obvious that the finished airfoil can not be accurate unless the original layout is extremely exact.

Aircraft Maintenance. Pg. 6

PROPELLER
Blade Hubs

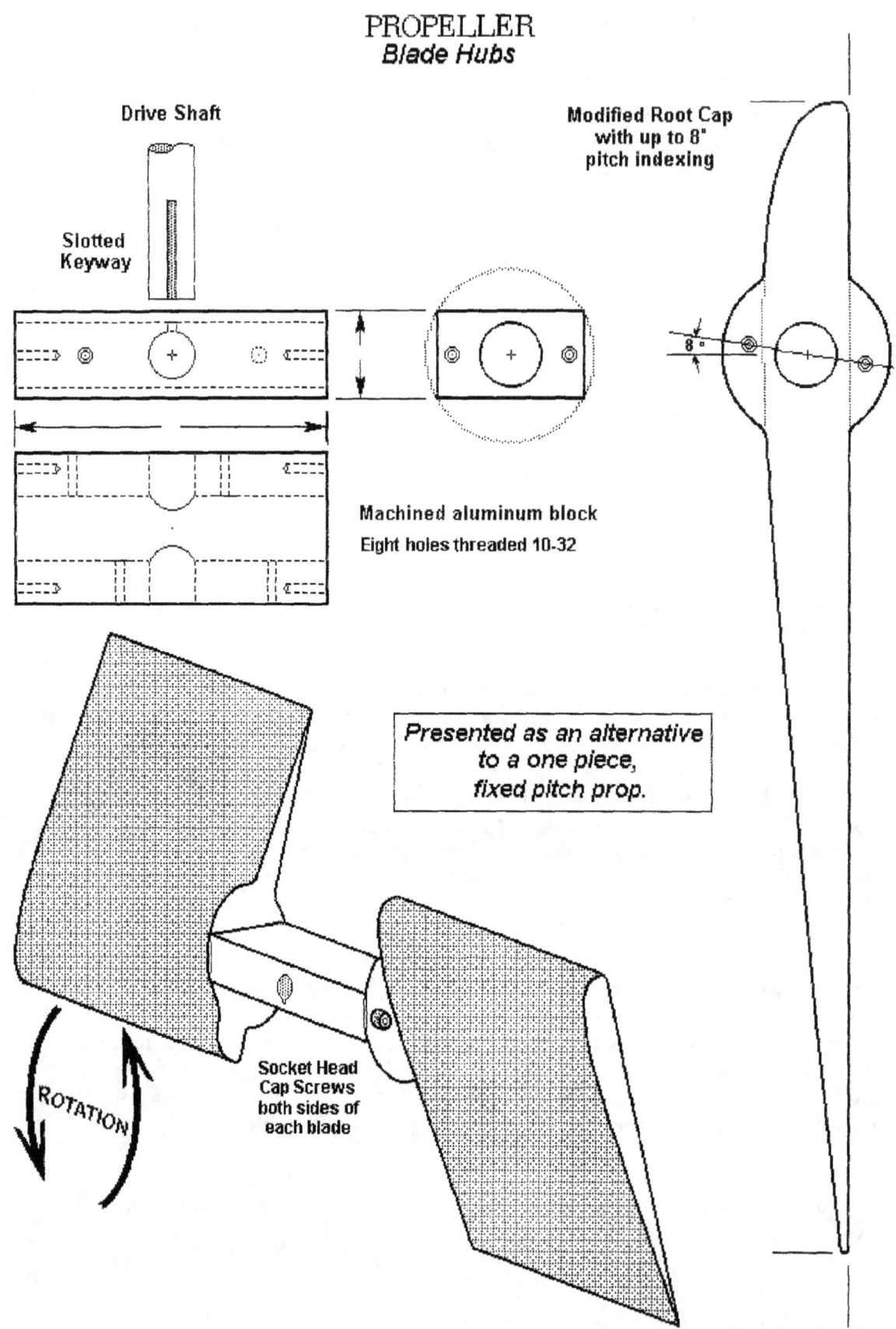

Drive Shaft

Slotted Keyway

Modified Root Cap with up to 8° pitch indexing

Machined aluminum block
Eight holes threaded 10-32

Presented as an alternative to a one piece, fixed pitch prop.

Socket Head Cap Screws both sides of each blade

ROTATION

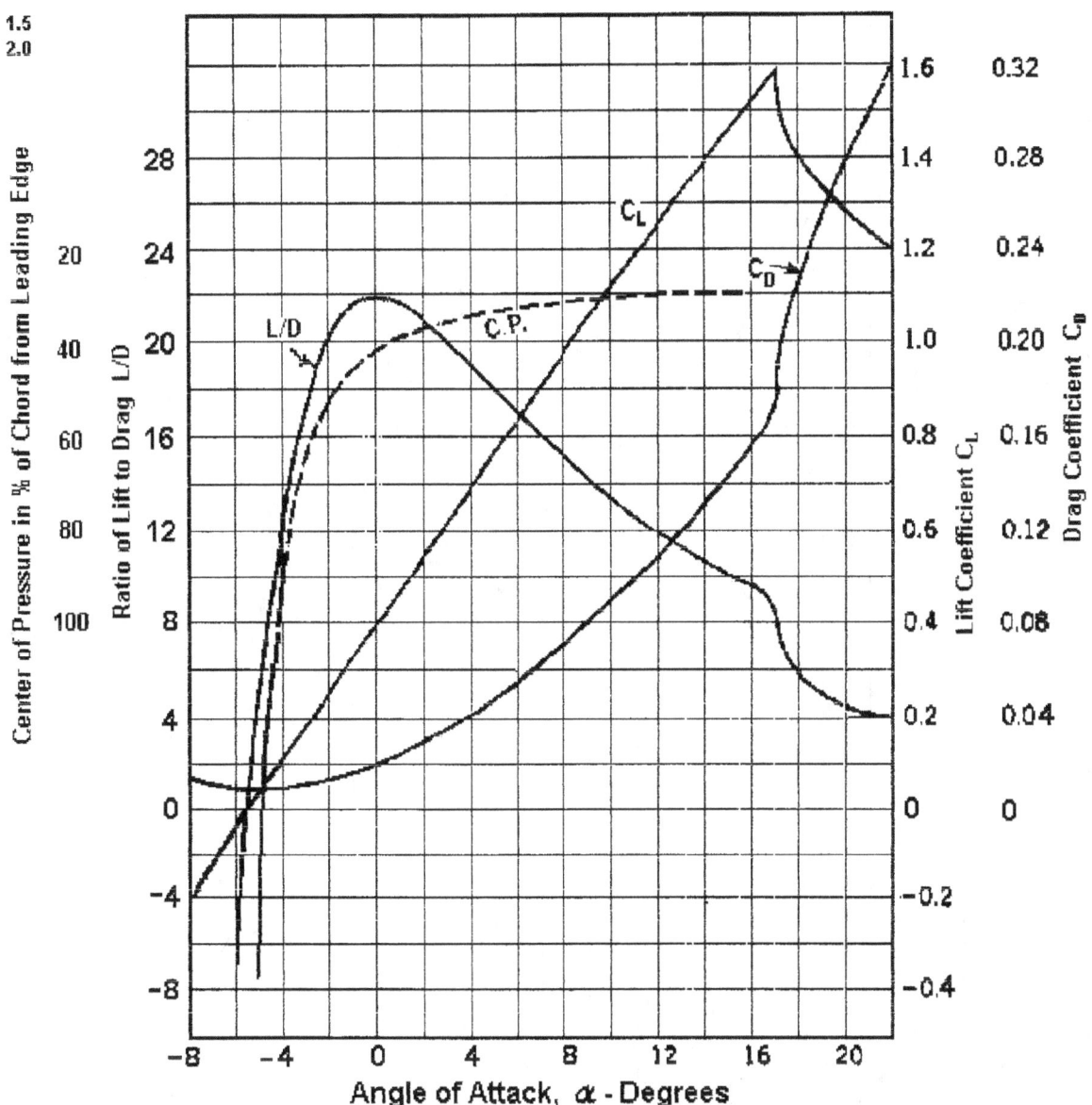

Clark-Y airfoil test data: aspect ratio 6; Reynolds number 6,000,000
(NACA Tech. Rept. 502)

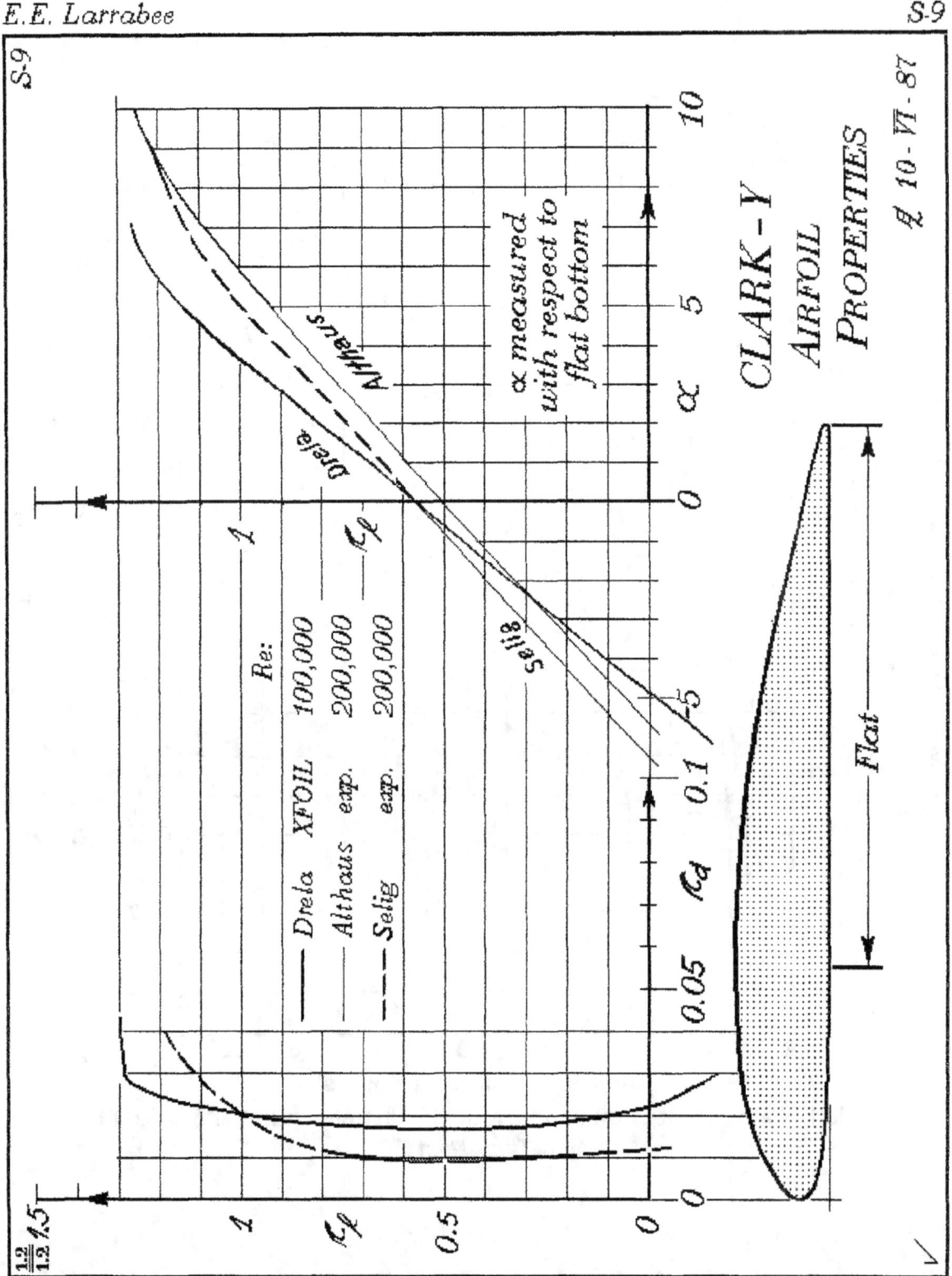

Wright Brothers
PROPELLERS

The low power end of engine driven prop scale may not be applicable to "muscle power" technology, but the problems that plagued the Wright Brothers may better enable us to see prop development in clearer light.

First, the Wrights had some success in building an internal combustion engine previous to 1901, but knew virtually nothing about props when they started. Their experiments began with gliders previous to 1901; engines and props came later. When they could not buy a ready made engine, they had to build one starting in December 1902. What they came up with in mid 1903 was about 9 horsepower (weighing 120 pounds).

As their engine experiments and testing progressed, only then did they give thought to propeller design. The Wrights had progressed well in their propeller work by the time of their first flight in June 1903. Their first props were made of three 1 1/8-inch spruce laminations glued together. Crudeness was the word; they were shaped by a hatchet and drawshave. They were 8.5 feet in length, 8 inches in width (total blade area 5.4 square feet), a 27 degree pitch, with the tips covered in canvas.

At the aircraft speed of 24 MPH, the two props turned at 330 RPM (tip speed 180 FPS) and produced a thrust of 90 pounds. As the sprocket ratio was about 3:1, the efficiency was close to 60%. It is interesting to note now that the chosen RPM was poorly chosen for the most favorable tip speed.

In 1904, four different propellers were used on the second airplane, including the 1903 propellers. Blades widths of 7.5, 8, 8.5, and 10 inches were used. The rotational speed was increased to 33:10, increasing the thrust to 185 pounds at 400 RPM. However, it was found that the added thrust caused an unwanted twist in the blades.

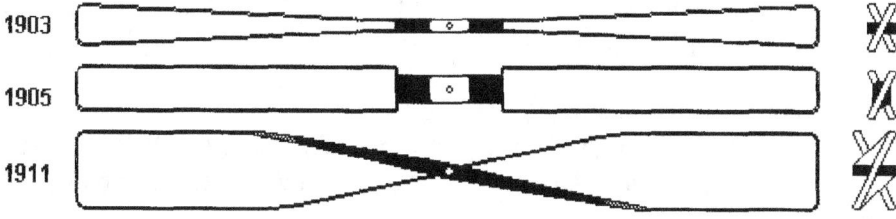

By 1905, they had done a complete redesign of their props. To reduce the twisting, a pie-shaped portion of the leading edge was removed resulting in a relative constant blade width for 30 percent of the distance from the tip toward the center. These propellers were called "bent end" type. From then on the Wrights continued the use of bent-ends with variations. Most propellers were 8.5 feet in diameter, though some were 9 feet in diameter. Blade widths ranged from 10 to 12 inches, pitch from 23 degrees to 27 degrees at the tips. Thrust on these props increased to 210 pounds at 450 RPM. Efficiency remained at 66% however. The outer two thirds of the blades were covered with light canvas glued to the front and back. The props were finished with aluminum paint, lacquered, and polished.

Propulsion Systems, p81-2

LOW-SPEED, HIGH EFFICIENCY
PROPELLERS

THE "GOSSAMER ALBATROSS" *AIRPLANE*

The ultralight airplane that Bryan Allen pedaled across the English Channel with June 19, 1979, is shown above. The propeller is configured as a "pusher" and is among the most efficient of their type ever designed. The airscrew of the *Gossamer Albatross* is 4.1 meters (13.5') in diameter, rotates at 95 r.p.m. and absorbs .8 hp. at takeoff and .25 h.p. in cruising flight. The propeller was computer-designed with the "Larrabee" algorithms to have minimum energy in its vortex wake and minimum friction loss in slightly climbing flight. The *Gossamer Albatross* was designed by AeroVironment Inc. of Pasadena, CA. Specs: Kevlar-carbon fiber, Eppler-193 airfoil.

THE "WHITE DWARF" *BLIMP*

The ultralight blimp Bryan Allen pedaled into the World record books in February 1985, is shown above. the propeller was also configured as a "pusher" but had an entirely different set of limitations. The frontal area of the blimp shape produced a much greater resistance to the air, and the rotating propeller had to be shorter, to clear the envelope and gondola.

Following are the basic design *criteria*:
    Target blimp speed: 15 mph (24.14 kph), 102.236 ft/sec.(31.162 mt/sec);
    Power input: 0.25 to 0.8 Hp.(192-614 watts); Cruise: 0.3 Hp.(230 watts).

Following are the basic design *limitations*:
    Prop Diam: 5.5 '(1.676 meters); Circumference: 17.279'(5.267 meters);
    Blimp max.frontal area: 189 Sq.Ft., Prismatic coefficient: .6683 (L=47.5').

A significant part of the computations, impossible to determine, was the "overall" drag of the aircraft as flown. Which includes all of the gondola, its contents, rudder, and the support cables. It does assume that the 6000 Cu.Ft. blimp is at equilibrium, and with a zero angle of attack. But since there were no previous records to beat, compromises were made to balance technology against economy. Specs: Straight tapered blades with 7.5 degree twist, spruce spar with hot-wire shaved foam interior, covered with one-mil Mylar. Airfoil is assumed as Clark-Y, but unverified.

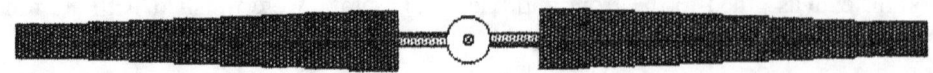

Source: Authors notes

Non Rocket Science
ENVELOPES

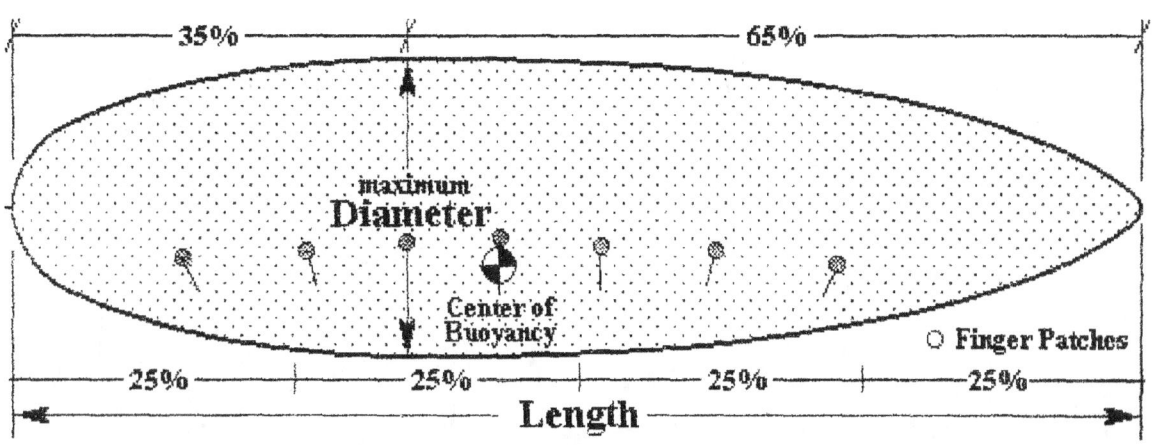

Targets
Blimp Target Speed: 15 MPH (22 Feet per Second)
Target Weight less than: 100 pounds (45.5 Kilos)

Limitations
Safe for sustained human flight; Gas tight, snag & tear resistant
Capable of at least 5 inflation cycles
Maximum Volume of 14,126 cubic feet (FAI Class BA1)
Envelope shape Drag Coefficient: .0136 (4:1 Fineness Ratio)

Compromises
Fabric Types: Urethane, or poly coated Dacron / Nylon
Indoor or Outdoor use (Ground Handling attachments)
Gas expansion control (pressure control)
Build yourself; Contract out (skill vs. quality product)

MUSCLE POWERED BLIMPS

MATHEMATICAL COMPUTATIONS

Mechanics of Spheroids

Problem: Finding the coordinates and abscissa for laying out a 75' long streamline shape having a fineness ratio of 4.0. **Fig. A** gives the *__proportions__ (k)* which may be used to lay out <u>any</u> airship shape. The relationship of the length to the thickness is called the fineness ratio. Thus the fineness ratio of 4 simply means that the spheroid called for is to be 4 times as long as it is thick. If the length is to be 75 feet then the thickness will be 1/4 of the length or 18.75 feet.

The vertical lines (y, radius) are the ordinate lines. To find at what point on the center line they are to be located from the nose, it is necessary to multiply the length (L) of the spheroid by the abscissa percentage. For example the first ordinate will be located .0125 percent of the length, or .0125 times 75 which equals .937' (11.25 inches) from the nose. **The location of <u>each</u> of the ordinate lines is determined in the same manner.**

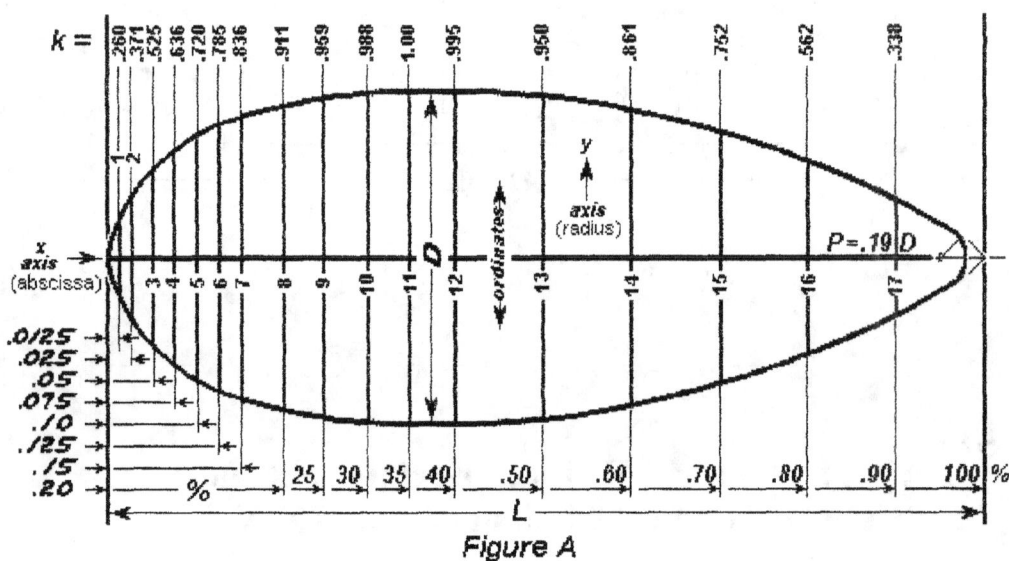

Figure A

To find the length of the ordinate lines multiply the max. diameter of the spheroid by the ordinate percentage. Thus the #1 ordinate will be 18.75 feet times .260 or 4.875'. However, this is the entire length of the ordinate so to find the location of the ends of the line from the center line it is necessary to divide the length by 2. Or to avoid having to do this each time, you can use 1/2 of the thickness, which in this case would be 2.43'.

For another example take ordinate # 11. This one is located at 35 per cent of the length or at 26.25' from the nose. As this is the widest point of the streamline, the ordinate percentage is 100 per cent. Thus the diameter of the ordinate line is 1.00 times 18.75 or 18.75'; and its radius half, or 9.375'.

NOTE: The quantity of coordinates is not be sufficient for smooth computor cutting.

AIRCRAFT SHEET METAL WORK p42-3.   R3°

MUSCLE POWERED BLIMPS

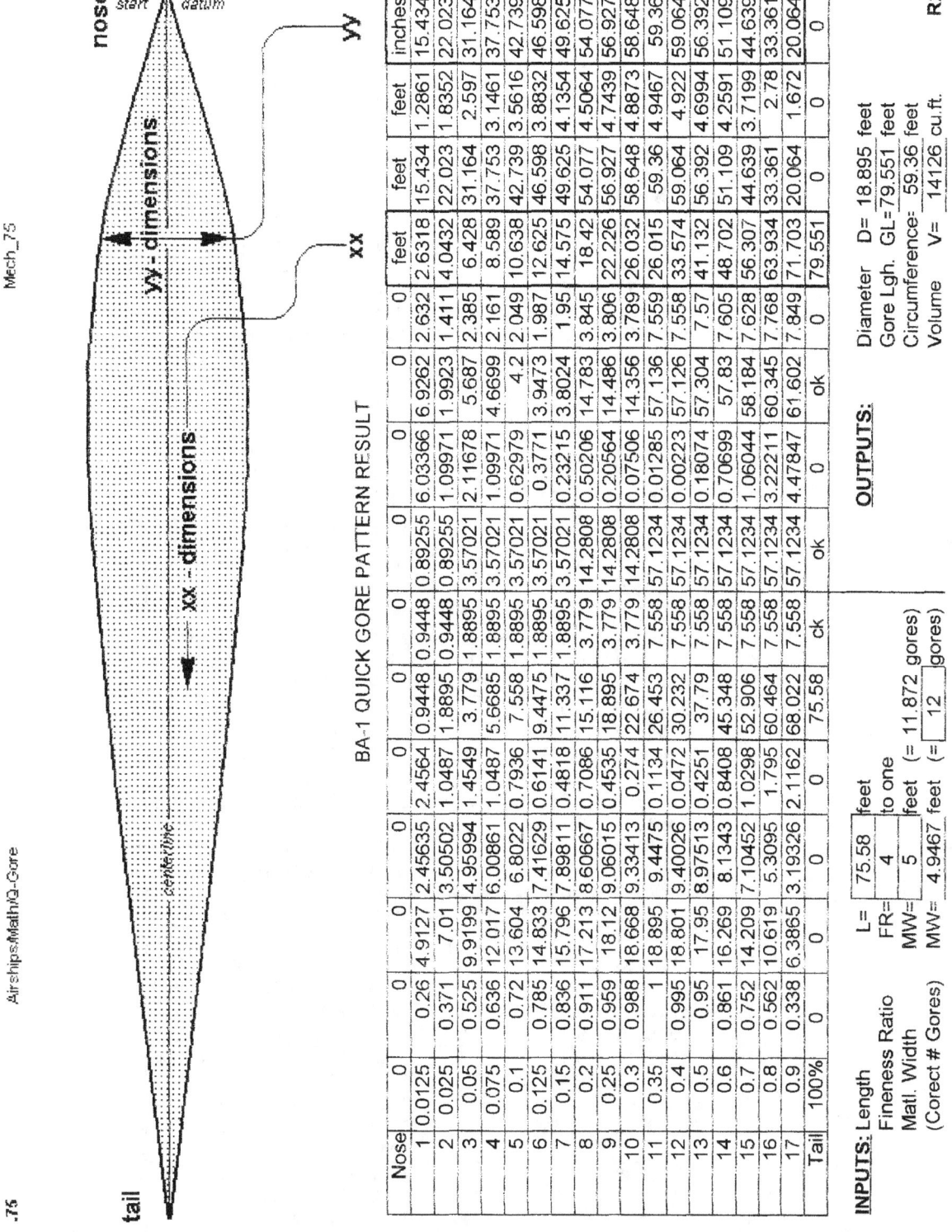

BA-1 QUICK GORE PATTERN RESULT

Nose															feet	feet	feet	inches
1	0.0125	0	0.26	4.9127	2.45635	2.4564	0.9448	0	0.9448	6.03366	0	6.9262	2.632	0	2.6318	15.434	1.2861	15.434
2	0.025	0	0.371	7.01	3.50502	1.0487	1.8895	0	0.89255	1.09971	0	1.9923	1.411	0	4.0432	22.023	1.8352	22.023
3	0.05	0	0.525	9.9199	4.95994	1.4549	3.779	0	0.89255	2.11678	0	5.687	2.385	0	6.428	31.164	2.597	31.164
4	0.075	0	0.636	12.017	6.00861	1.0487	1.8895	0	3.57021	1.09971	0	4.6699	2.161	0	8.589	37.753	3.1461	37.753
5	0.1	0	0.72	13.604	6.8022	0.7936	1.8895	0	3.57021	0.62979	0	4.2	2.049	0	10.638	42.739	3.5616	42.739
6	0.125	0	0.785	14.833	7.41629	0.6141	1.8895	0	3.57021	0.3771	0	3.9473	1.987	0	12.625	46.598	3.8832	46.598
7	0.15	0	0.836	15.796	7.89811	0.4818	1.8895	0	3.57021	0.23215	0	3.8024	1.95	0	14.575	49.625	4.1354	49.625
8	0.2	0	0.911	17.213	8.60667	0.7086	3.779	0	14.2808	0.50206	0	14.783	3.845	0	18.42	54.077	4.5064	54.077
9	0.25	0	0.959	18.12	9.06015	0.4535	3.779	0	14.2808	0.20564	0	14.486	3.806	0	22.226	56.927	4.7439	56.927
10	0.3	0	0.988	18.668	9.33413	0.274	3.779	0	14.2808	0.07506	0	14.356	3.789	0	26.032	58.648	4.8873	58.648
11	0.35	0	1	18.895	9.4475	0.1134	7.558	0	57.1234	0.01285	0	57.136	7.559	0	26.015	59.36	4.9467	59.36
12	0.4	0	0.995	18.801	9.40026	0.0472	7.558	0	57.1234	0.00223	0	57.126	7.558	0	33.574	59.064	4.922	59.064
13	0.5	0	0.95	17.95	8.97513	0.4251	7.558	0	57.1234	0.18074	0	57.304	7.57	0	41.132	56.392	4.6994	56.392
14	0.6	0	0.861	16.269	8.1343	0.8408	7.558	0	57.1234	0.70699	0	57.83	7.605	0	48.702	51.109	4.2591	51.109
15	0.7	0	0.752	14.209	7.10452	1.0298	7.558	0	57.1234	1.06044	0	58.184	7.628	0	56.307	44.639	3.7199	44.639
16	0.8	0	0.562	10.619	5.3095	1.795	7.558	0	57.1234	3.22211	0	60.345	7.768	0	63.934	33.361	2.78	33.361
17	0.9	0	0.338	6.3865	3.19326	2.1162	7.558	0	57.1234	4.47847	0	61.602	7.849	0	71.703	20.064	1.672	20.064
Tail	100%	0	0	0	0	0	0	0	0	0	0	0	0	0	79.551	0	0	0
						75.58		ck	ok		ok							

INPUTS: Length L= 75.58 feet
Fineness Ratio FR= 4 to one
Matl. Width MW= 5 feet (= 11.872 gores)
(Corect # Gores) MW= 4.9467 feet (= 12 gores)

OUTPUTS:
Diameter D= 18.895 feet
Gore Lgh. GL= 79.551 feet
Circumference= 59.36 feet
Volume V= 14126 cu.ft.

R2

MUSCLE POWERED BLIMPS

HULL COORDINATES (Max.Dia.@ 40% Length) Page 1 of 2

INPUTS:	Length	L=	82.10954	feet	OUTPUTS	Diameter D=	18.246564
Fineness		FR=	4.5	to one	Max.circumference	C=	57.323407
Matl.width		MW=	5	feet	Volume (cu.ft)	V=	14000
# of Gores		N=	12	pieces		XX	YY
X-Sta-%	X-Distance	Y-Radius	Cum.Surf	Cum.Vol.		Gore Pos.	Gore Width
0	feet	feet	sq.ft.	cu.ft.		0	0
1	0.8210954	2.118	11	12		2.2715901	1.1089848
2	1.6421908	2.972	26	34		3.4562899	1.5561392
3	2.4632862	3.611	45	68		4.4967315	1.8907196
4	9.8531448	4.137	66	112		5.471859	2.1661332
5	4.105477	4.588	90	166		6.4086614	2.4022768
6	4.9265724	4.985	116	231		7.3206957	2.610146
7	5.7476678	5.34	143	304		8.2152474	2.796024
8	6.5687632	5.662	172	387		9.0972234	2.9646232
9	7.3898586	5.955	203	478		9.9690299	3.118038
10	8.210954	6.224	235	578		10.833066	3.2588864
11	9.0320494	6.471	269	686		11.690508	3.3882156
12	9.8531448	6.701	303	802		12.543208	3.5086436
13	10.67424	6.913	339	925		13.391231	3.6196468
14	11.495336	7.11	376	1056		14.235628	3.722796
15	12.316431	7.293	413	1193		15.076869	3.8186148
16	13.137526	7.464	452	1337		15.915581	3.9081504
17	13.958622	7.623	491	1487		16.75193	3.9914028
18	14.779717	7.771	531	1642		17.586257	4.0688956
19	15.600813	7.908	572	1804		18.418703	4.1406288
20	16.421908	8.036	613	1970		19.249715	4.2076496
21	17.243003	8.155	655	2142		20.079389	4.269958
22	18.064099	8.265	698	2318		20.90782	4.327554
23	18.885194	8.367	741	2499		21.735227	4.3809612
24	19.70629	8.462	785	2683		22.5618	4.4307032
25	20.527385	8.548	829	2872		23.387386	4.4757328
26	21.34848	8.628	873	3064		24.21237	4.5176208
27	22.169576	8.701	918	3259		25.036704	4.5558436
28	22.990671	8.767	964	3457		25.860448	4.5904012
29	23.811767	8.827	1009	3658		26.683732	4.6218172
30	24.632862	8.881	1055	3862		27.506601	4.6500916
31	25.453957	8.929	1101	4067		28.329099	4.6752244
32	26.275053	8.971	1147	4275		29.151267	4.6972156
33	27.096148	9.008	1194	4484		29.973196	4.7165888
34	27.917244	9.039	1240	4695		30.794876	4.7328204
35	28.738339	9.065	1287	4907		31.616383	4.746434
36	29.559434	9.087	1334	5120		32.437774	4.7579532
37	30.38053	9.103	1381	5334		33.259025	4.7663308
38	31.201625	9.114	1428	5548		34.080194	4.7720904
39	32.022721	9.121	1475	5763		34.901319	4.7757556
40	32.843816	9.123	1522	5977		35.722417	4.7768028
41	33.664911	9.121	1569	6192		36.543534	4.7757556
42	34.486007	9.115	1616	6406		37.364703	4.772614
43	35.307102	9.104	1663	6620		38.185936	4.7668544
44	36.128198	9.089	1710	6833		39.007251	4.7590004
45	36.949293	9.07	1757	7045		39.828668	4.749052
46	37.770388	9.047	1804	7257		40.650208	4.7370092
47	38.591484	9.02	1850	7466		41.471851	4.722872
48	39.412579	8.99	1896	7675		42.293692	4.707164
49	40.233675	8.955	1943	7882		43.115666	4.688838
50	41.05477	8.917	1989	8087		43.937785	4.6689412

R2

MUSCLE POWERED BLIMPS

HULL COORDINATES (Max.Dia.@ 40% Length) Page 2 of 2

X-Sta-%	X-Distance	Y-Radius	Cum.Surf.	Cum.Vol.	XX Gore Pos.	YY Gore Width
51	41.875865	8.876	2034	8290	44.760112	4.6474736
52	42.696961	8.831	2080	8491	45.582668	4.6239116
53	43.518056	8.782	2125	8690	46.405409	4.5982552
54	44.339152	8.73	2170	8887	47.228344	4.571028
55	45.160247	8.675	2215	9081	48.051485	4.54223
56	45.981342	8.617	2260	9272	48.874918	4.5118612
57	46.802438	8.555	2304	9461	49.698582	4.479398
58	47.623533	8.49	2348	9647	50.522489	4.445364
59	48.444629	8.422	2391	9830	51.346648	4.4097592
60	49.265724	8.351	2434	10010	52.171071	4.3725836
61	50.086819	8.277	2477	10187	52.995769	4.3338372
62	50.907915	8.2	2510	10360	53.820753	4.29352
63	51.72901	8.12	2561	10530	54.645932	4.251632
64	52.550106	8.038	2602	10697	55.471519	4.2086968
65	53.371201	7.952	2643	10860	56.297317	4.1636672
66	54.192296	7.864	2684	11020	57.12344	4.1175904
67	55.013392	7.773	2724	11176	57.949898	4.0699428
68	55.834487	7.679	2764	11328	58.776586	4.0207244
69	56.655583	7.583	2803	11476	59.603628	3.9704588
70	57.476678	7.484	2841	11620	60.430912	3.9186224
71	58.297773	7.383	2880	11761	61.258568	3.8657388
72	59.118869	7.279	2917	11898	62.086606	3.8112844
73	59.939964	7.172	2954	12030	62.914773	3.7552592
74	60.76106	7.064	2.991	12159	63.743472	3.6987104
75	61.582155	6.952	3.026	12284	64.572307	3.6400672
76	62.40325	6.839	3052	12404	65.401556	3.5809004
77	63.224346	6.723	3096	12521	66.231087	3.5201628
78	64.045441	6.605	3130	12634	67.06105	3.458378
79	64.866537	6.484	3164	12742	67.891307	3.3950224
80	65.687632	6.361	3197	12846	68.713106	3.3306196
81	66.508727	6.327	3229	12947	69.562392	3.3128172
82	67.329823	6.11	3260	13043	70.393559	3.199196
83	68.150918	5.981	3291	13135	71.225197	3.1316516
84	68.972014	5.849	3321	13224	72.056994	3.0625364
85	69.793109	5.716	3351	13308	72.889113	2.9928976
86	70.614204	5.581	3380	13388	73.722058	2.9222116
87	71.4353	5.441	3408	13465	74.555685	2.8489076
88	72.256395	5.297	3435	13537	75.390914	2.7735092
89	73.077491	5.144	3462	13605	76.228032	2.6933984
90	73.898586	4.981	3487	13669	77.067988	2.6080516
91	74.719681	4.804	3512	13729	77.911921	2.5153744
92	75.540777	4.609	3536	13783	78.761207	2.4132724
93	76.361872	4.392	3559	13833	79.618361	2.2996512
94	77.182968	4.146	3580	13878	80.487184	2.1708456
95	78.004063	3.862	3600	13916	81.373612	2.0221432
96	78.825158	3.528	3618	13948	82.2896	1.8472608
97	79.646254	3.122	3634	13973	83.258838	1.6346792
98	80.467349	2.607	3648	13991	84.351559	1.3650252
99	81.288445	1.886	3657	14000	86.408545	0.9875096
100	82.10954	0	3657	14000	87.229641	0

Gore width	GW=	4.7769506	feet	
Ctr.Buoyancy	CB=	36.54	feet	
Gore Lgh.		87.229641	feet	
Solidity SR=		0.6521	units	

Helium Tight
ULTRA LIGHT FABRICS

<u>**POLYESTER, 150 Dinier, Polyurathane Coated, 4.5 oz/sq.yd**</u>. (#1 choice)
This is the lightest, and best conventional fabric you can buy. It can be thermal welded, glued, or sewn. But if you want a good shape, with the strongest-toughest envelope for its weight, this is absolutely the material. It is good for 5-years if kept outside, or 10 years inside a hangar. If you intend to deflate it, and keep it in a box during the Winter, it should be good for 3 Summers. It has great scuff resistance, but the more times it is creased, the more pinholes that will appear. It is best thermal welded with the "Tape technique".

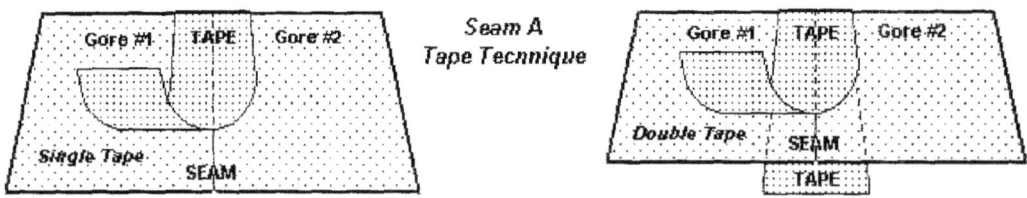

The best quality fabric is available from **LAMCOTEC**, whose address is to be found in the list of suppliers.

<u>**POLYURATHANE, 9-mil (.009") thickness, Unpigmented**</u>. (best one shot, throw away)
This is the lightest fabric available that is still considered safe to fly. It will give you a beautiful shape, and is capable of withstanding a 100% overinflation. It also can be thermal welded or glued. It can NOT be sewn (and gluing is not recommended). It has a long life if not abused, but it won't take much scuffing on cement surfaces. It will punctire fairly easily on a sharp object, but since it is a resilient fabric, even a large tear will not propagate. It should also be capabable of a 3-season shelf life. It is best thermal welded with the "edge technique".

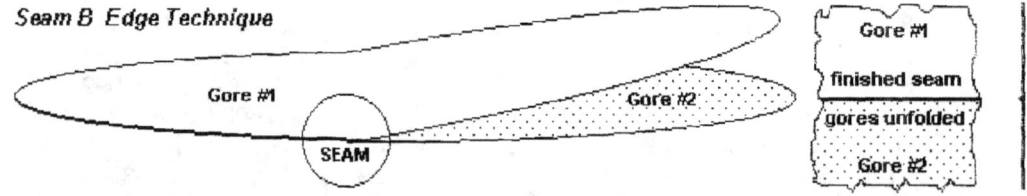

The closest *Urethane fabric-film Supplier* can be found on the following page.

A complete line of Thermal Sealing machines are available from ***Vinyl Welders Inc***. of Navarre, Ohio. They can also supply a list of nearby customers to weld it for you.

Muscle Power Materials

Working with POLYURETHANE FILM

Why use Polyurethane Film?

Because it is cheaper, holds Helium well, easy to work with, AND very light. It not only stretches, but is fairly resilient and moderately UV proof. It won't take much abrasion, or any snagging, but then with reasonable care in handling, it will easily last a summer (or two). I consider it as a great "throw-away" envelope. Its unique property is in its stretch-resiliency.

Why is stretch-resiliency important? (See chart)

That allows a small blimp to adjust its volume as the temperature changes, without the need of ballonets. The stretch-limit on a blimp is in how far you can distort it and still have it return to its original length when you release the tension (internal pressure). If you exceed that limit, the volume will keep increasing and require more & more helium or it will start to look "saggy". Obviously, you will need more ballast until it develops a leak. Or worse yet, ruptures.

So how is the resiliency determined? (See drawing)

I recommend that you test a sample of each roll you buy. No "rocket science", just a simple test can be very accurate. What you are trying to determine is the point at which the stretch is permanent by carefully taking a series of measurements under load, and no load. When it won't return in a reasonable time (5 minutes), STOP. Remember, each thickness will be different. And any wild temperature change during the testing will skew the results. Just taking a blimp out of a shady hangar into sunlight can change the temperature 5 degrees. So the test results will be important for the temperature range you expect to operate in.

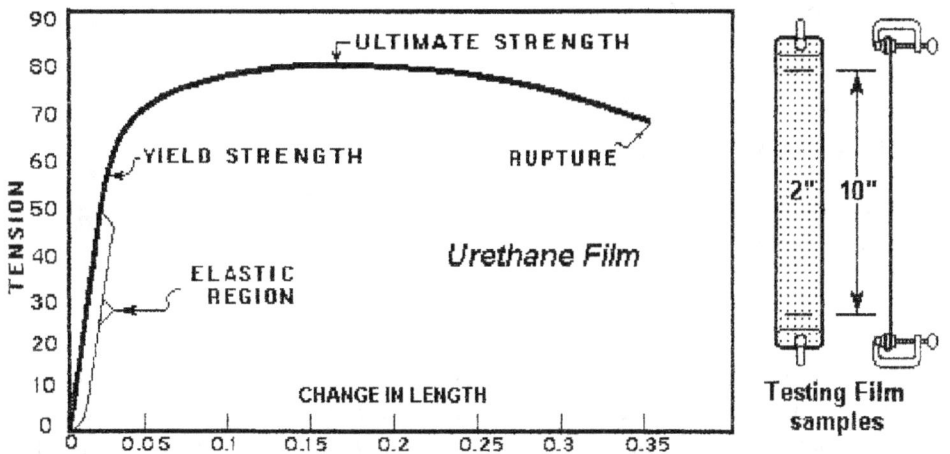

Figure 1. Points of interest on the elastic curve. Note that the elasticity is proportional to its limit and is hard to recognize thereafter.
(units not accurate for a specific sample)

How do I make realistic use of the resulting data?

You are going to put 2 horizontal lines, 10" apart, on the belly of the envelope. And periodically measure the distance between the lines. After a little experience you can recognize an approaching limit at a glance. Your "wake-up call" is when you have trouble getting back to the ground. So pay attention, 5 degrees temp. change changes your lift 10%.

MUSCLE-POWER BLIMPS
(FABRIC SUPPLIERS)

FILM, Urethane Plastic
(9-mil, Inflateable, Thermal weldable)

AMERICAN National Mfg.　　　800 / 854-6295
363 American Circle
Corona, CA　　　91720

BECKWITH-BEMIS Inc.　　　800 / 667-8469
54 West Dane Street
Beverly, MA　　　01915

DEARFIELD Urethane　　　800 / 644-0116
Post Office Box 186-A　　　413 / 665-7016
So.Dearfield, MA　　　01373

EHOB, Inc.　　　800 / 966-3462
250 North Belmont
Indianapolis, IN　　　4622?

EVANS Industries　　　888 / 775-4716
606 Walnut Ave. NE
Canton, OH　　　44702

FILMSTAR Inc.　　　513 / 459-9776
4503 US Route 42
Mason, OH　　　45040

MANN Industries　　　508 / 626-0518
225 Arlington Street POB 689
Framingham, MA　　　01704

THORODIN, Inc.　　　N/P
5500Central Ave, Ste.203
Boulder, CO　　　80301

NOTE: Inspect & thouroghly test material samples before you buy. Thickenss should be uniform .009 inches thick. It should stretch 100% and return to size within 45 minutes. A 2 inch cut should withstand a 10 pound perpendicular pull without propagating. A 3 inch raveled strip should take 300% stretch without breaking.

See your Thomas Register. Get quality guarantee before price.　　　R:11/2/98

MUSCLE POWERED BLIMPS

R: 9/23/98

FABRIC & WEBBING
SUPPLIERS

AIRCRAFT SPRUCE Co. 201 Truslow Av.(POB 424) Fullerton, CA 92632	800 / 824-1830 Great A/C source	FABRICS, Ceconite Fin coverings dopes and tapes
BALLY Mills 23-25 North 7th Street Bally, PA 19503	800/845-2201 f)610/845-8013 quality products	FABRICS, polyester Mill Spec (MS) Mfgr. highly respected
LAMCOTEC Inc.(RECOMENDED) 152 Bethany Rd. (POB-279) Monson, MA 01057	413/267-4808 f)413/267-5265 Rick Henderson	FABRICS, polyester Ask for sample Made to your specs
LOWRY Aircraft Supply 2311 E.Artesia Blvd. Long Beach, CA 90?81	310/531-8134 hard to find items great source	FABRICS, surplus surplus & seconds inspect closely
MANN-Tech Industries 225 Arlington Way Farmingham, MA 01701	617/879-6366 well advertized but unknown	FABRICS, polyester Dacron & more
MICROSEAL Industries 610-T East 36th Street Patterson, NJ 07509	201/523-0704 well known quality work	FABRIC coatings Polyester laminating
ORCON Corporation 1570 Atlantic Street Union City, CA 94587	800/228-2781 exotic scrims heat sealing	POLY FILMS Reinforced films great helium barriers
PERFORMANCE Textiles 3917 Liberty Road Greensboro, NC 27406-6109	910/275-5800 Fax:275-8866 good source	FABRICS, polyester Dacron & more large quanities
STANDARD Brands Paints 620 Sepulveda Blvd. Van Nuys, CA 91401	818/786-1381 most Calif. cities ck yellow pages	FIBERGLASS cloth & resin Bondo, glues, paint
VINYL WELDERS Inc. 4220 Alabama Ave. SW Navarre, OH 44662	330/833-6739 fax: 833-2828 BEST source	Thermal Welders Complete range www.vinylwelders.com
UNITED Textile (Unitex) 5175 Commerce Drive. Baldwin Park, CA 90012	818/962-6281 Retail/wholesale good source	FABRICS/hardware WEBBING, canvas H-66 fabric glue<------

NOT recommended: *Gentek*

Specialty Items; See your Thomas Register. Get quality guarantee before price.

MUSCLE POWERED BLIMPS

Thermal Welding machines

Retractable Grappling Hook

The Retractable Grappling Hook (RGH) is designed to provide a high strength, compact, lightweight anchor for rope or cable. The device is machined or forged from high strength steel or metallic alloy and it provides the user with a means of anchoring a rappel rope, rescue line or climbing rope when a "tie-off" method cannot be used. It also provides a means to breach trip-wire and tilt-rod type detonators.

The Retractable Grappling Hook is specifically designed for easy carrying and portability. It weighs 1.17 pounds. Its novel mechanical design provides extra strength through interlocking of its hook elements when deployed.

The prototype RGH has been pull tested to deformation in it's deployed (open) configuration. Forces were placed on a single hook element and load shackle to simulate worst case conditions. The shaft began to bend at 2000 pounds. No evidence of deformation was observed in the hook element.

The RBH is a product of the MobD S&T program and a patent application is being processed. Details or a demonstration are available upon request.

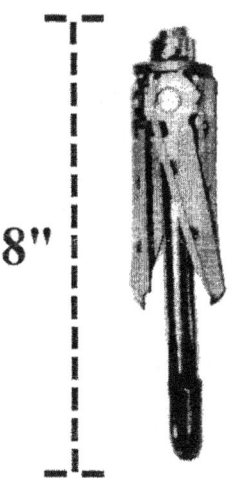

MUSCLE POWERED BLIMPS

Muscle Power

chapter 4

Ergonomics

Getting Fit, Keeping Fit

Fitness

Improving with
ENDURANCE TRAINING

ROLLERS are really bicycle treadmills. For racing, they are geared through cables to a centralized dial roller is pointed a different color and is connected to an arrow of matching color on the dial so that his speed can be followed and compared directly to the rate of other riders.

Roller races are generally short—seldom more than two miles. The riders perspire heavily, and there is no wind to evaporate the perspiration. But with no wind resistance, the racers attain high speeds, often close to 60 miles per hour.

Rollers were designed originally for winter training, and they are often used for warming up before track races. Rollers are considerably more difficult to ride than the rood, for they are quite narrow (about 16 inches). Rollers help develop rapid pedal action.

BICYCLE ERGOMETERS are refined bicycle exercisers. They hone have built-in mechanisms for measuring the work exerted. Ergometers have proved useful to athletes in training, for they provide an accurate means of determining the amount of effort spent in exercise and, equally important, give a measure of progress. Their use is often prescribed by physicians or by physio-therapists for patients whose physical effort is needed in recovery from leg injuries. Ergometers are especially valuable in determining heart rate during exercise.

Often an air-collecting device is used in conjunction with the ergometer so that the body's ability to consume oxygen can be measured at the same time energy is being spent. this combination is particularly valuable in diagnosing the body's total cardiovascular condition and its recuperative powers.

The ergometer's work load is determined either by electrical resistance loads or by friction braking systems. By changing dials or levers, the work load can be varied when either greater or less strain is desired. The work output measurements are given as electrical watts or at calories.

The ergometer illustrated here has an electrically measured work output and a built in calorie tachometer. It gives a constant reading of the riders pulse *rate* while the machine is being ridden.

PAYOFF; Incredible speeds can be attained on a bicycle when there is little wind resistance. Tandems, triplets (3 riders), and up to quints (5 riders) were used to set the pace for speedy racers in the early days of racing. The most renowned of those who established records, Charles "Mile-a-Minute" Murphy who, in 1899, rode one mile behind a train in 57 4/5 seconds.

Over the years this record was increased steadily until the magic 100 miles per hour speed was passed in 1942, by Alfred Letourner, who rode a mile in 33.05 seconds—a speed of 108.92 miles per hour. Jose Meiffret achieved 127.98 miles per hour in 1962, and Dr. Al Abbott pedaled 138.67 miles per hour in 1973. In all of these the racers rode behind some kind of pacing machine.

A distance record was set by Hubert Opperman, an Australian, who cycled 860 miles 367 yards in 24 hours.

Bicycling, Golden Press, p115.

R

MUSCLE POWERED BLIMPS

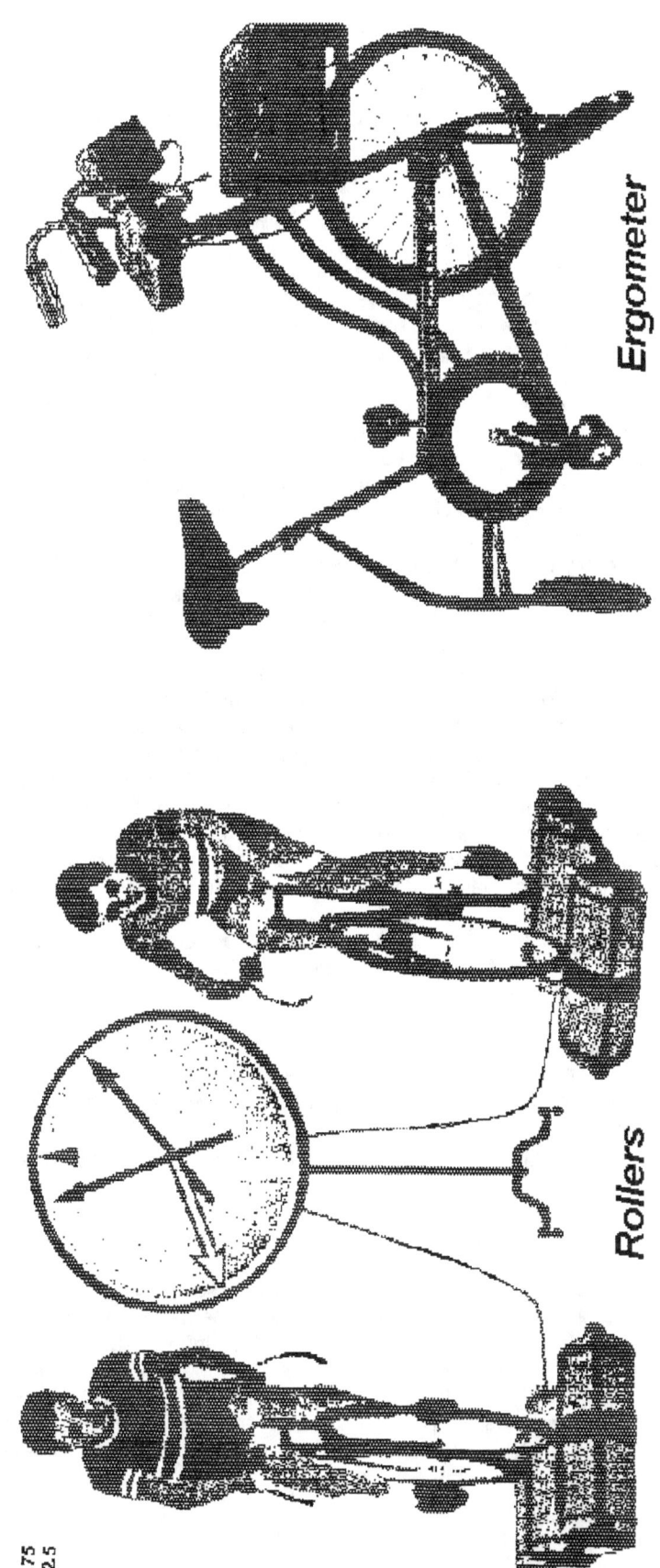

Pilot fitness

Feel the Power
(Strength Training gives you the edge)
by Jerry Davis

It happens every weekend: You are in a group of racers, closing in on the finish. The whole race has been a battle, back and forth; all five in the group seem to have good legs under them. The final climb before the finish is on the horizon, and it's a stiff quarter-mile uphill climb. As the group rolls to its base, one rider drops directly into the granny gear, while the others do their best to keep pace. Suddenly, one rider jumps hard and goes 10 feet off the front Thirty seconds later, he's over the top and out of sight. And without ever looking back, he continues to go strong, catching the group ahead, and winning the sprint finish. Due to his seemingly fresh legs, he gained six places in the last two miles. And the best part? That rider was *you*.

On the drive home, your riding partners ask, 'What's up with you?" They remember doing the same event last year when you finished mid-pack. This year, you are in the top five experts. It is easy to re-live the suffering you went through last year, while this year you were able to stand up and hammer. No matter how hard you tried last year, you just couldn't erase the deep pain and frustration of losing places each time you hit a power climb."

Trying to identify what created the monster of power today, you can look back at your training log. It holds the key: During the off season your strength training never disappeared. You have maintained weight work throughout the year, setting a personal record high of power at your lactate threshold just a few weeks ago.

Many athletes train for mountain-bike races as if they were racing a time trial—going against the course and clock while trying to keep a steady intensity throughout. And in some ways, that's exactly what's happening. However, there are few mountain courses which allow racers to settle into a steady pace, even on long climbs. Once a racer does settle into a steady pace, an uphill and back extension. With a good warm-up of 20 minutes, two sets of these 11 exercises with a minute rest in between, and another 20 minutes of cooling down, you should be in and out of the weight room in less than one hour.

Don't forget to do your "power intervals." Pick your favorite, short (one minute), steep climb, attack it in a big gear (at least two gears higher than normal), and keep power in the climb until you are up and over the top. Take three minutes to recover after the effort, spinning easily at a high cadence in the power of your legs that you (more than 90 rpms), and then repeat. Do only six to 10 of these intervals. Many more may tap into your race-day reserves.

After every intense workout and race effort eat or drink at least 100 grams of carbohydrates within 30-45 minutes afterward, in order to speed recovery. It doesn't need to be the most high-tech sports drink. A bagel or banana will work. The important thing is to get the carbos into your system while your body is in the "super-compensation" mode, grade will increase or decrease dramatically, and the racer has to power up, drop back, power up, drop back ... again and again.

Every corner of a criterium is similar; decelerate for the turn, accelerate back up to speed afterward. Think about how many times you go through that sequence in the course of a crit. Imagine having so much confidence in the power of your legs that you can jump hard out of every turn, and power up every climb. Can you now see yourself gaining positions throughout a race?

Everyone has either seen or been a part of such a scenario. The questions are: Which rider are you? The one with deep power? Or one of the other racers who struggled to the line? Do you find yourself wanting the power you had in the spring, which seems to elude you mid-season? The answer is strength training.

Strength training can help clear "blocked" legs through the dynamics of weight lifting, and, by elongating the muscles, can create greater range of motion. And with an effective warm-up and cool down time, athletes will be able to get back into their high-intensity training more quickly after racing.

VELONEWS, July 27, 1998 p36-38.

MUSCLE POWERED BLIMPS

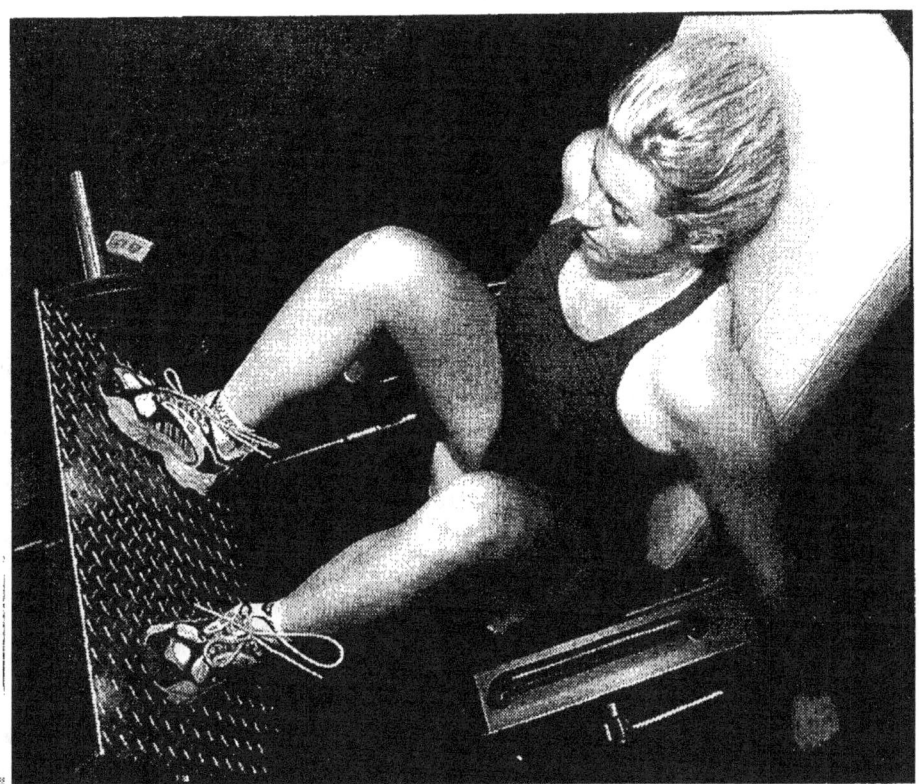

PUMP IT UP Don't let the racing season turn your schedule upside down. Keep lifting.

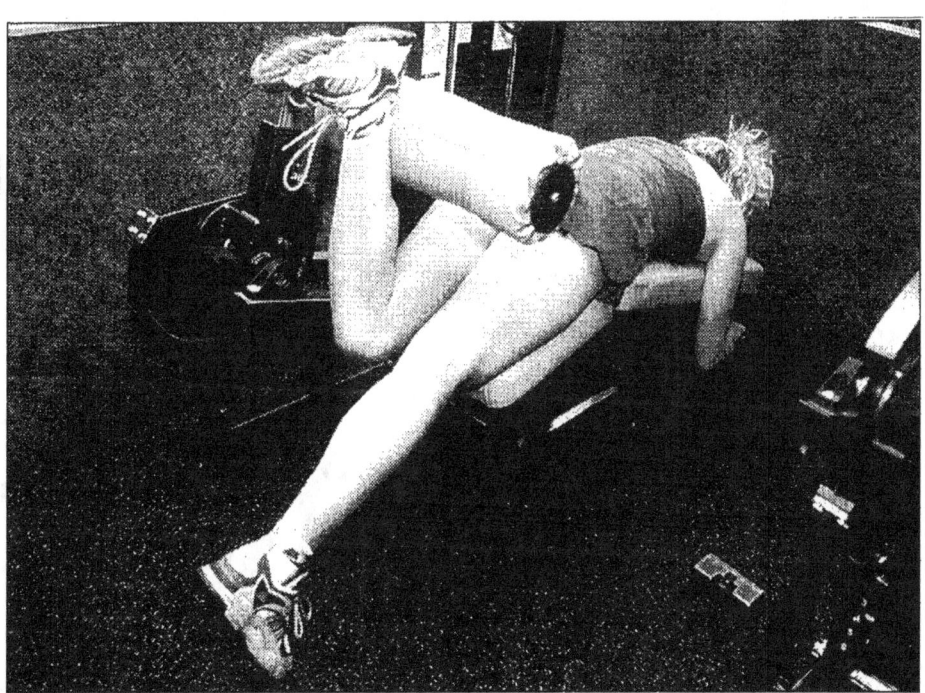

60 MINUTES With the proper workout, you can get a leg up on the competition with an hour in the weight room.

Pilot fitness

Here is a list of strength training exercises and other tips that will keep power in your legs throughout the summer:

- √ Stay in the weight room. Keep your workouts to two sets, hitting fatigue around 25 repetitions. Keep the exercises simple: leg press, leg extension, leg curl, calf raise, seated row, bench press, upright rows, arm curl, triceps extension, abdominal crunch which occurs immediately following moderate to intense exercise.
- √ In order to benefit from all of this hard work, you must have proper race nutrition and hydration. Consume 300-400 calories per hour when racing, and drink, drink, drink. Without the proper fuel, your body cannot produce the necessary energy (power) on demand. Eat, drink or slurp—whatever it takes—your energy replacement every 15 - 20 minutes and follow that with 66 ounces of water.

To keep your body in check while strength training, make sure to stretch. Maintaining proper body alignment will help eliminate chronic aches and pains. You may want to enroll in a yoga class in which you can learn the subtleties of proper and effective stretches. And if you get in the habit of stretching everyday, you will find that chronic injury can be deterred by relieving the tension in troublesome spots.

So, get off your saddle, and dedicate a few hours per week to improving fitness and power through strength training. You will be able to use this cross trainirg tool as an advantage over your competition, and it will build confidence when you need to tap into your deep power reserves. Spending time pushing weights will enable you to be the athlete who schools the competition, rather than being fee one wearing a dunce cap for a helmet

Jerry Davis is a USA Cycling elite level coach, and an American Council on Exercise "Gold" certified personal trainees He can be reached at 360/83S7677, or mtbcoach@2aol.com.

VELONEWS, July 27, 1998 p36-38.

MUSCLE POWERED BLIMPS

Pilot fitness

Measuring the Power
(SRM-Strength Recording Measurement)
by Matt Mantell

With 58 km. left in the 1997 Amstel Gold Race, Bjarne Riis and seven other riders accelerated ahead of the peloton. As the escapees extended their lead on the flat stretch of road, Riis sensed he could pedal even faster. After al, both legs corfirmed the Telekom rider's observation. As did the numbers—power 450 watts, pulse rate 160 bpm, speed 50 kph—that appeared on the SRM computer affixed to his bike's handlebars.

So, like a Great Dane fleeing from a pack of annoying dachshunds, the 1996 Tour de France champion bounded away for this World Cup event's *arrive* banner. Over the next 40km, studded with liffle Limburg hills, Riis distanced himself from his less powerful coconspirators. And by the time he crossed the finishing stripe, the Herning, Denmark native had a 46second gap on the second-placed cyclist. For the record, the last rider to post a classic victory inside of a season of winning the Tour was Bernard Hinault at the 1980 Liege-Bastogne-Liege.

Yet, the Frenchman, unlike his Scandinavian counterpart, neither raced nor trained with the SRM Power Measuring System. This high-tech gizrno, the brainchild of Germany's Ulrich Schoberer, wasn't introduced to the bunch until 1986. And since then, numerous elite riders—Greg LeMond, Andy Hampsten, Chris Boardman, Miguel Indurain, Jan Ullrich, just to name a few—have utilized the $5000 piece of equipment. However, it is used only during practice sessions by most riders.

Which, of course, Riis does, too. I train alone a lot, so for me, it's the best riding partner you can have," enthused the normally stoic Riis. "But you have to know how to analyze the data. I know how to do that It's obvious...."

Perhaps. But there's no doubt that Riis is one of the few riders to use the SRM during a race. Other cyclists, citing the unit's additional weight, remain reluctant to ferry around the 570 gram cranks. Riis, though, mininizes the significance of this extra ballast in competition. Especiallywhen the tradeoff is obtaining valuable information offered by the strain gauges housed in the chainrings.

It would be very interesting to see how much power I would make in the (1998) Tour," said Riis, who lastyear finished a disappointing seventh. Walter Godefroot (Telekom's directeur sportif) doesn't want me to because fee SRM isn't one of our sponsors. But I have used the SRM in the Tour's time trials. And, of course, at the Amstel Gold Race."

To get the lowdown on Riis's World Cup tour de force, his first *classic* win, *VeloNews* asked Schoberer to analyze fee rider's SRM race graph. So, put on your thinking caps, you road scholars, and gear up for some mental exercise.

Note that the chart has four lines. In order tom top to bottom, they represent heart rate, speed, power and cadence.

Yet, speed doesn't wear on the glaph's final hour. At that juncture, Riis suffered a puncture and his new wheel didn't have a sensor. What's rnore, over the same time period, the cyclists's pulse rate wasn't accurately recorded. During the race's final 60 minutes, the television motor camera that dogged the Dane's spectacular breakaway disrupted the heartrate monitor's transmission.

These glitches notwithstanding, the SRM generated many intrusive statistics. "Looking at Riis's total file: we can see from the beginning of the race the speed is very high," said Schoberer. This is probably because the road is flat and Riis is riding in the peloton. When, the speed goes down a little. So does the power. Maybe, because of the wind. But toward the end of the race, the power rises higher and higher." (The horizontal axis displays time and distance [in km].)

So, what conclusions can be drawn from this overview of Riis's numbers?

"The race really starts after four hours of riding," said Schoberer.

"And that Riis can still go 400 watts." Most notably, during an ascent.

In the right half of the screen, you'll notice a hill. Riis's speed sharply drops down. But his power rises to nearly 500 watts.

VELONEWS, July 27, 1998 p36-38.

Pilot fitness
cont.

"In the Amstel Gold Race, the course has a lot of small bergs, or hills, where you need to climb for more than two minutes," Schoberer pointed out. "In a World Cup race, if you want to win, you must have the ability to do two be three minutes at close to 400 watts. You can see that again and again, Riis has the ability to do that in the hills. And that after a race of 260km, he can still produce 400 watts. Only the top riders can do this."

Plus sustain a solo escape for 40km averaging both 390 watts and 85.5 revolutions per minute.

"These numbers are not so much," noted Schoberer, before he added, but you have to remember the race was 230km old. And that Riis had already climbed about 20 hills at 530 watts."

Then Schoberer quickly added, "Heart rate is absolutely not important. That's why it's all right that this parameter doesn't appear on Riis's graph."

Wait a second. This assertion requires some justification, particularly when so many cyclists rely on their pulse rates to guide and gauge their fitness.

'When you have a motorcar, the horsepower and the aerodynamics are important," explained Schoberer. But the revolutions of the motor is not so important. In cycling, you can have one rider whose heart rate is 170 (beats per minute) and another with 180. Who's better? You can't tell.

"But let's say, you have two riders who are the same weight. One can do 500 watts and the other 300. From this, it's easy to determine which rider is better. With power (watts), you can see how good someone is and how they make progress in training."

To further support these contentions, the scientist trotted out Riis's pertinent data. "During the winter, his maximum heart is, maybe, 190 (bpm)," said Schoberer. "And in summer, it's also 190. But his maximum power in winter, for four minutes, is 380 (watts) and in summer it's 500. You can see a difference of 120 watts—but the *heart rate* is the same."

Still, in Schoberer's opinion, power output must be put in the context of how much a cyclist weighs. Especially during a stage race.

"At the Tour de France, to be with the best riders, you need to be able to do 480 watts when you have a body weight of 70 kilograms," he said. 'Ten years ago, maybe 460 watts were enough. (Marco) Pantani uses the SRM system. For him, with 55 Kilos, 400 watts is enough. When you have the weight of 80 kilos, like (Mario) Cipollini, then you have to produce 600 watts. Last year, when Ullrich weighed 74 kilos, he could ride up to 500 watts

In the hills, weight is the most important. But on the flats, the only thing that's crucial is aerodynamics. Look at Boardman. His weight is like Ullrich's. But Boardman—he trains with the SRM system—and has very good aerodynamics. He did his hour record, I think, at 450 watts. And he won the world's two years ago by doing 60km with only 530 watts. Other riders would need 600 or 650 watts "

After uttering these thought-provoking observations, Schoberer offered some words of caution. This data only has meaning for people who have an SRM," he admitted. "Because if you don't have the system, you won't know what it means to do a hill with 400 watts. You will only **improve** in cycling when you make almost every training ride an attempt to produce more power with the same body weight.

"Power is the key to winning a race."

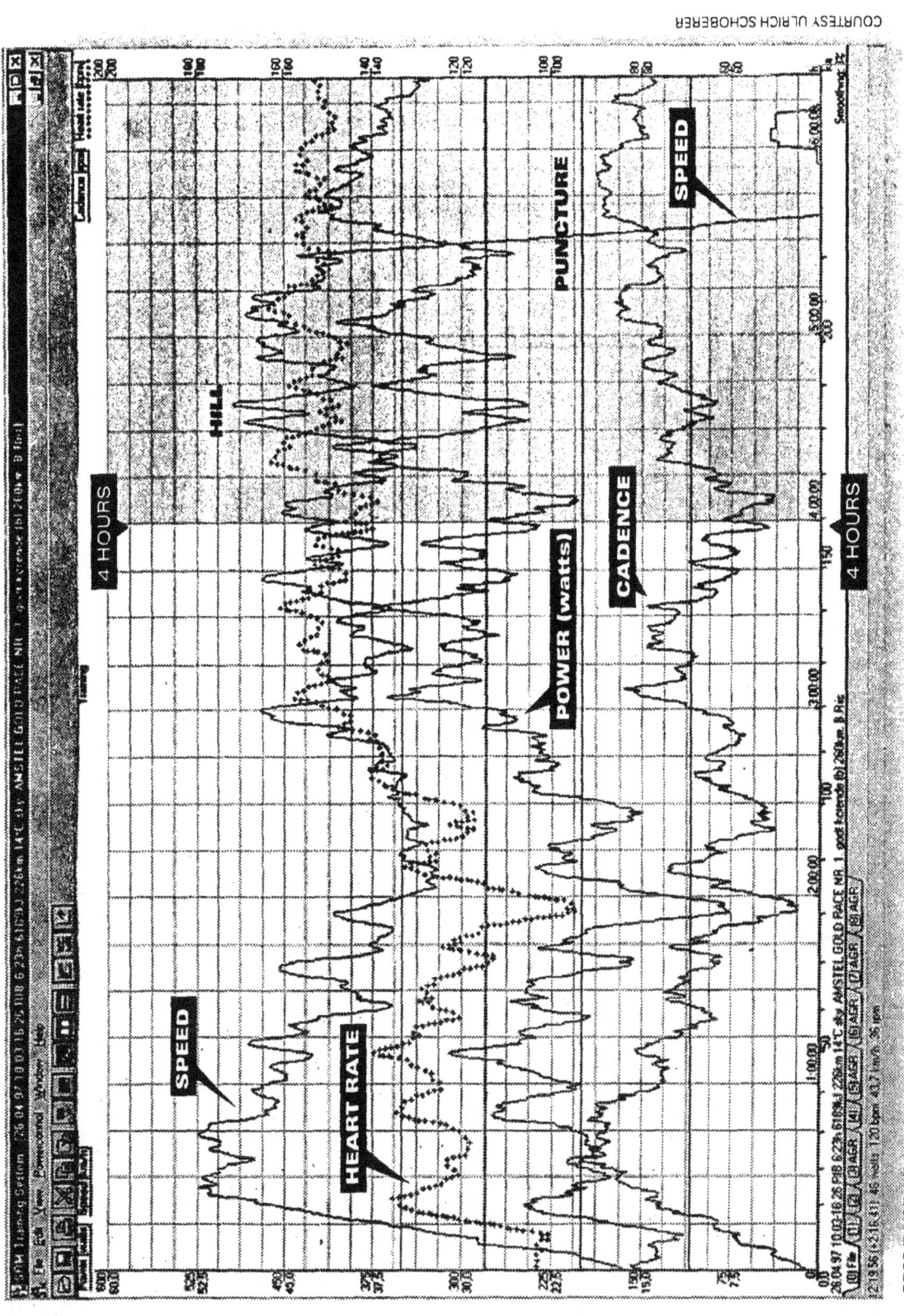

SRM RACE GRAPH: The chart above, generated by the SRM software, displays Bjarne Riis's power, heart rate, cadence and speed for the 1997 Amstel Gold Race.

ROCKET SCIENCE

for
PEDAL POWER

Prepared by
Robert J. Recks

PROPELLER THEORY

DEFINITIONS

The following definitions, with the exception of those marked *T*. are taken from National Advisory Committee for Aeronautics reports. They should be studied and learned to save time in using this book.

aerodynamics—The branch of dynamics which treats of the motion of air and other gaseous fluids, and of the forces on solids in motion relative to such fluids.
airfoil—Any surface designed to be projected through the air in order to produce a useful dynamic reaction.
airfoil section (or profile)—A cross-section of an airfoil made by a plane parallel to a specific reference plane. A line perpendicular to this plane is called the axis of the airfoil.
airplane—A mechanically driven aircraft, heavier than air, fitted with fixed wings, and supported by the dynamic action of the air.
air speed—The speed of an aircraft relative to the air. Its symbol is V.
angle:
    critical angle—The angle of attack at which the flow about an airfoil changes abruptly, with corresponding abrupt changes in the lift and drag.
downwash angle The angle through which an air stream is deflected by lifting surface of an airplane. It is measured in a plane parallel to the plane of symmetry.
    effective helix angle—The angle of a helix described by a particular point on a propeller blade as the airplane moves forward through air otherwise undisturbed.
    zero lift angle—The angle of attack of an airfoil when the lift is zero.
angle of attack—The acute angle between the chord of an airfoil and its direction of motion relative to the air. (This definition may be extended to other bodies than airfoils.)
angle of incidence —(Pertains to an airplane wing setting).
    angle of pitch—The angle of propeller blade (usually blade angle)—The acute angle between the chord of a propeller section and a plane perpendicular to the axis of rotation of the propeller.
angle of stabilizer setting—(Pertains to an airplane wing setting).
aspect ratio—The ratio of the span to the mean chord of an airfoil, i.e., the ratio of the square of the span to the area of an airfoil.
aspect ratio of propeller blade—Half the ratio of the propeller diameter to the maximum blade width.
blade back—The cambered Side of a propeller blade, corresponding to the upper surface of an airfoil.
blade face—The surface of a propeller blade which corresponds to the lower surface of an airfoil.
camber—The convexity or rise of the curve of an airfoil section from its chord, usually expressed as the ratio of the maximum departure of the curve from the chord to the length of the chord. Top camber refers to the upper surface of an airfoil and bottom camber to the lower surface. Mean camber is the mean of these two.

The Aircraft Propeller P1-12

PROPELLER THEORY

DEFINITIONS

center of pressure— (c.p) of an airfoil section—The point in the chord of an airfoil section, prolonged if necessary, which is the intersection of the chord and the line of action of the resultant air force.

chord of an airfoil section—The line of a straight edge brought into contact with the lower surface of the section at two points; in the case of an airfoil having double convex camber, the straight line joining the leading and trailing edges. (These edges may be defined for this purpose as the two points in the section which are farthest apart.) The line joining the leading and trailing edges should be used also in those cases in which the lower surface is convex except for a short flat portion. The method used for determining the chord should always be explicitly stated for those sections with regard to which ambiguity seems likely to arise.

chord length (C)—The length of the projection of the airfoil section on its chord.

drag (D)—The component parallel to the relative wind of the total air force on the aircraft or airfoil. The absolute drag coefficient is **Cd** as defined by the equation **Cd =D/qs**, in which **D** is the drag, **q** is the impact pressure **(l/2pV2)** and **s** is the effective area of the surface upon which the air force acts. In the case of an airplane, that part of the drag due to the wings is called wing drag; that due to the rest of the airplane is called structural drag or parasitic resistance.

induced drag—That portion of the airfoil drag induced by, or resulting from, the generation of lift.

profile drag—That portion of the airfoil drag which is due to friction and turbulence in the fluid and which would be absent in a non-viscous fluid.

horsepower of an engine, maximum—The maximum horsepower which an engine can develop.

horsepower of an engine, rated—The average horsepower developed by an engine of a given type in passing the standard 50 hour endurance test.

indraft, inflow— The flow of air from in front of the propeller into the blades.

leading edge (also Called entering edge)—The foremost edge of an airfoil or propeller blade.

lift (L)—That component of the total air force on an aircraft or airfoil which in perpendicular to the relative wind and in the plane of symmetry. It must be specified whether this applies to a complex- aircraft or parts thereof. (In the case of an airship this is often called dynamic lift.) The absolute lift coefficient is **Cl** undefined by the equation **Cl = L/qs,** in which **L** is the lift, **q** is the dynamic pressure **(l/2 pV)**, and **s** is the effective area of the surface upon which the air force acts.

Momentum Theory—In this theory the propeller is treated as a disk producing a uniformly distributed thrust created by a pressure differential between the front and back of the disk. A stream tube encloses the affected streamlines which are considered continuous through the disk. Air is assumed a perfect fluid with no viscosity or Impressibility effects.

pitch of a propeller:

    aerodynamic pitch— distance a propeller would have to advance in one evolution in order that the torque might be zero.

The Aircraft Propeller P1-12

PROPELLER THEORY

DEFINITIONS

effective pitch—The distance an aircraft advances along its flight path for one revolution of the propeller.

geometrical pitch—The distance an element of the propeller would advance in one revolution if it were moving along a helix of a slope equal to its blade angle.

mean geometrical pitch—The mean of the geometrical pitches of the several elements.

standard pitch (also nominal pitch)—The geometrical pitch taken at two-thirds of the radius.

virtual pitch (also experimental mean pitch)—The distance a propeller would have to advance in one revolution in order that there might be no thrust.

pitch ratio—The ratio of the pitch to the diameter.

pitch speed—The product of the mean geometrical pitch by the number of revolutions per unit of time, the forward speed made if there was no slip.

propeller—Any device for propelling a craft through a fluid such as water or air; especially a device having blades which when mounted on a power driven shaft, produce thrust by their action on the fluid.

adjustable propeller—A propeller whose blades are so attached to the hub that the pitch may be changed while the propeller is at rest.

automatic propeller—A propeller whose blades are attached to a mechanism that automatically sets them at their optimum pitch for various flight conditions.

contra-propeller—A second propeller mounted in the immediate rear of a first propeller. The second propeller rotates in the opposite direction to that of the first propeller to overcome the rotation of the airplane due to propeller torque. (T)

controllable propeller—A propeller whose blades are so mounted that the pitch may be changed while the propeller is rotating.

left-hand propeller—A propeller rotating anti-clockwise as viewed from the cockpit. (T)

propeller, area projected—The total area in the plane perpendicular to the propeller shaft swept by the propeller root of the blade This portion is usually taken as extending .02 of the maximum radius from the axis of the shaft.

propeller axis—The axis on which the propeller rotates being the engine or propeller shaft.

propeller balancing stand—A stand on which knife edges are placed and upon which rest a mandrel or arbor, which is placed through the hub of an assembled propeller, and the assembly checked for static balance. Tile propeller should remain motionless in any position. (T)

propeller, blade area—The area of the blade face exclusive of the boss and root, i.e.., of a portion which is usually taken as extending 0.2 of the maximum radius from the axis of the shaft.

propeller boss—The central part of the propeller in which the hub is located.

propeller, camber ratio—The ratio of the maximum thickness of a propeller section to its chord.

The Aircraft Propeller P1-12

PROPELLER THEORY

DEFINITIONS

propeller coefficients—Mathematical expression of the characteristics of a propeller, such as thrust, drag, etc., when the V/ND is a constant as the lift and drag coefficient of an airfoil at a set angle of attack.

propeller diameter (blade disk)—The diameter of a circle made by the blade tips rotating. (T)

propeller disk areas—The area swept by a propeller in rotation being a circle having a diameter equal to that of the propeller.

propeller efficiency—The ratio of the thrust horsepower developed to the engine horsepower delivered to the propeller. Output/Input.

propeller interference—The amount by which the thrust and torque of a propeller are changed lay the modification of the airflow in the slip stream produced by bodies placed near the propeller, such as engine, radiator, etc.

propeller load curve—A curve representing the engine power necessary to drive a given propeller at various speeds. The power required varies approximately as the cube of the speed in r.p.m. provided the *V/ND* ratio remains constant.

propeller, race rotation—The rotation produced by the action of the propeller, of the stream of air passing through or influenced by the propeller.

propeller rake—The angle which the line joining the centroids of the sections of a propeller blade makes at the axis with a plane perpendicular to the axis.

propeller reduction gear—A propeller shaft and gear assembly driven off the engine shaft. Due to the gearing, the propeller rotates snore slowly than the engine shaft rotation, thus reducing tip speeds. (T)

propeller root—That part of the propeller blade near the hub.

propeller section—A cross-section of a propeller blade made at any point by a plane parallel to the axis of rotation of the propeller and tangent at the centroid of the section to an arc drawn with the axis of rotation as its center.

propeller thrust—The component parallel to the propeller axis of the total air force on the propeller. Its symbol is **T**.

propeller torque The moment produced on the propeller by the engine shaft.

propeller width ratio, total—The product of the blade width ratio, at the point of maximum blade width by the number of blades.

pusher propeller—A propeller mounted to the rear of the engine or propeller shaft. (It is usually behind the wing or nacelle).

right-hand propeller—A propeller rotating clockwise as viewed from the cockpit. (T)

tandem propellers—One propeller mounted as a tractor on the foremost engine and facing the direction of flight, the second propeller being mounted on the aft engine facing the tail of the airplane.

tractor propeller—A propeller mounted on the forward end of the engine or propeller shaft. (It is usually forward of the fuselage or wing nacelle.)

slip—The difference between the geometrical pitch and the effective pitch. It may be expressed as a percentage or linear value.

The Aircraft Propeller P1-12

MUSCLE POWERED BLIMPS

PROPELLER THEORY

DEFINITIONS

slip function—The ratio of speed of advance through undisturbed air to the product of the propeller diameter by the number of revolutions in unit time, i.e., *V/ND*. The slip function is the primary factor controlling propeller performance. It is "pi" times ratio of forward speed to the tip speed of the propeller.

slip stream—The stream of air driven astern by the propeller. The indraft *is* sometimes included also.

span: (b), airfoil span—The lateral dimension of an airfoil: its dimension perpendicular to its chord.

stalling—The condition of an airfoil which, from any cause, has lost the effective control of the airsflow.

streamline flow—Flow past a solid body without any discontinuity in the pressure of the velocity distribution.

streamline form—A solid body which produces Streamline flow.

thrust:

    effective thrust—The net driving force delivered by a propeller when mounted on an airplane: the actual thrust given by a propeller as mounted on an airplane, minus any increased resistance of the airplane produced by the action of the propeller.

    static thrust—The thrust developed by a propeller when rotating at a fixed point.

tipping, propellers—sheet metal (or equivalent) protective covering on the blade of a propeller near the tip, extending a short distance along the trailing edge and a considerable distance along the leading edge.

tractor airplane—airplane with the propeller or propellers forward of the main supporting surfaces.

trailing edge—The rearmost edge of an airfoil or rotating propeller blade.

warp—To change the form of a wing by twisting it. Warping is sometimes used to maintain the lateral equilibrium of an airplane. NOT to be confused with twist on a propeller blade).

wash (also called the wake in the general case for any solid body). The disturbance in the air produced by the passage of an airfoil.

wash-in—Permanent warping of the wing which results in an increase in the angle of attack near the tip.

washout—Permanent warping of the wing which results in a decrease in the angle of attack near the tip.

wind, relative—The motion of air with reference to a body: Its motion as observed by a person at rest upon the body.

wind tunnel—A chamber for testing airfoils.

wing loading—The gross weight of an airplane, fully loaded, divided by the area of the supporting surfaces. The area used in computing the wing loading should include ailerons, but not the stabilizer of elevators.

zero lift line—A line through the trailing edge of an airfoil section parallel to the direction of the wind when the lift is zero.

The Aircraft Propeller P1-12

AIRFOIL DESCRIPTION
Reading the Numbers

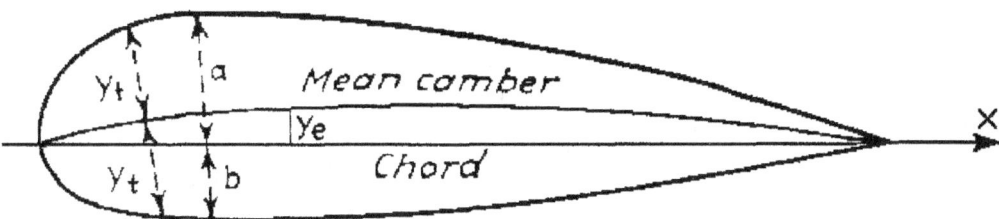

A typical airfoil section or profile is shown above with the chord of length "c" between the leading and trailing edge. Half way between the upper and lower surfaces of the airfoil runs the **Mean Chamber Line** (Fig.A), usually of parabolic shape. The thickness of the airfoil is measured at right angles to the mean camber line. Maximum camber is usually located at 30 per cent from the leading edge. The exraordinary lifting power of cambered wings as compared with flat boards is one of the most gratifying results of aerodynamic research.

The angle between chord and relative wind is the **Angle of Attack, "a"**, of the wing. The angle between the propeller axis or direction of thrust and relative wind is known as the **Angle of Incidence**. The lift of the airfoil increases with increasing **"a"** up to a certain maximum value, above which the lift decreases again.

Airfoil profiles are denoted by 4-index symbols, like (6315). This symbol means: The maximum camber of the wing is 6 per cent of the chord; it is located at 3 tenths of **"c"** (or 30 per cent) from the leading edge, the maximum thickness is 15 per cent of **"c"**. A number of profiles with their 4-digit symbols are shown in Fig. B. If the second digit (3 tenths) is replaced by two digits (30 per cent) one arrives at the 5-digit symbols used lately in the reports of the N.A.C.A., *i.e.*, National Advisory Committee for Aeronautics.

An airfoil section may also be characterized by the maximum upper camber, **"a"**, above the chord and the maximum lower camber below the. The upper camber is always positive, whereas the lower camber may be "positive," to produce a convex-concave wing, or "negative," yielding a double convex wing.

The Physics of Flight, p29-30

AIRFOIL PROFILES
National Advisory Committee on Aeronautics

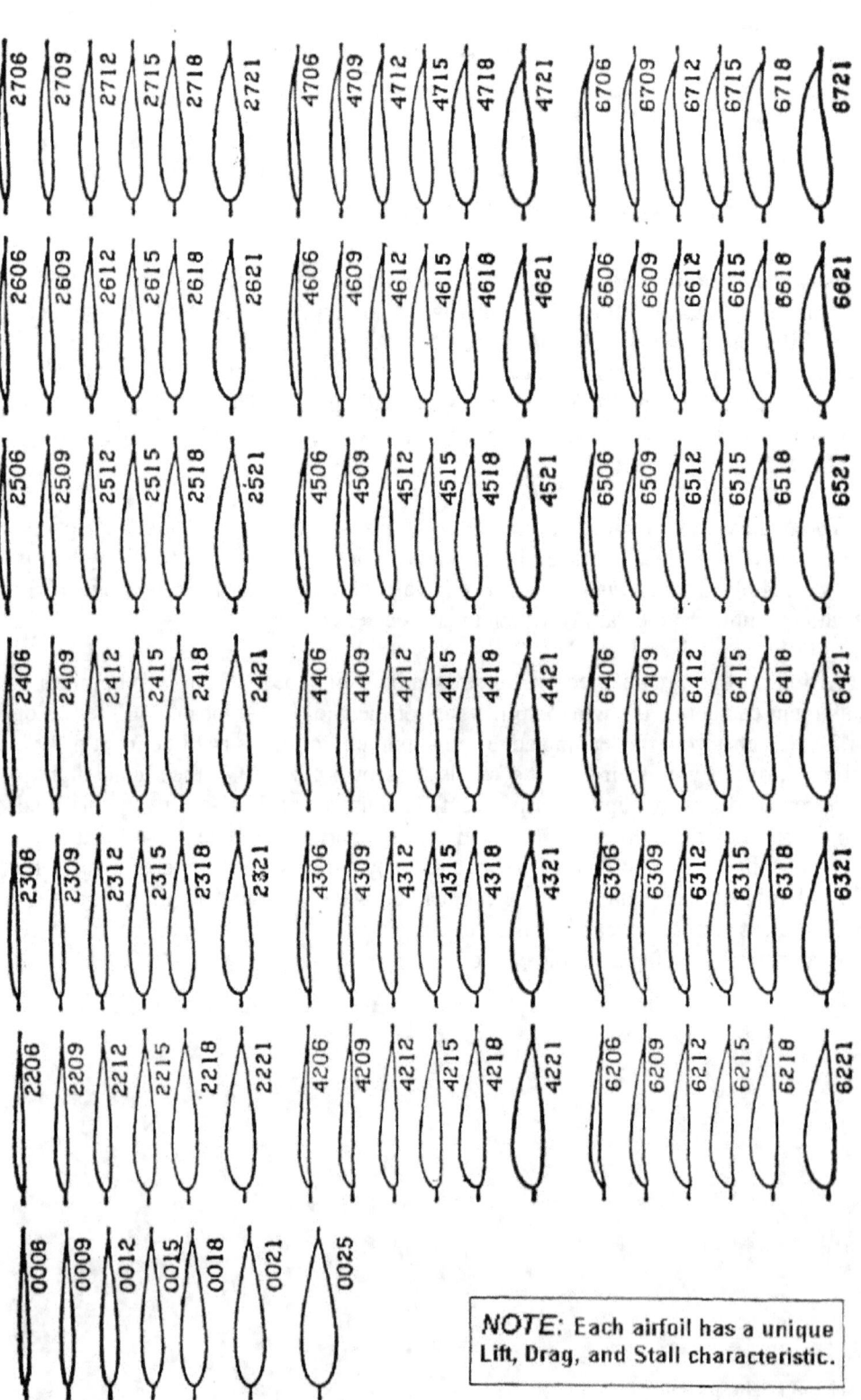

NOTE: Each airfoil has a unique Lift, Drag, and Stall characteristic.

Fig. B Sample NACA Airfoils

Lifting-Line Formulation for Counter-Rotating Rotors
Mark Drela, MIT Aero-Astro Dept.

Summary

This section describes the analytical formulation used for the design and analysis of counter-rotating rotors. It is based on a lifting-line representation of the rotor blade together with a general semi-free wake method used to describe the induced velocities. Simple two-dimensional profile drag characteristics are used to account for viscous losses.

Although this is a simpler model than could be constructed with a general 3-D vortex-lattice or panel method, it is more than adequate for accurate prediction of the aerodynamic performance of the rotors. Even the general 3-D formulations would have to make the same time-averaging assumptions for the unsteady counter-rotating flow as the present method, and hence would not be more sophisticated or more accurate in this regard. The chief advantage of the present method is that it is extremely fast computationally and is has simple inputs, making it ideal for interactive design work.

The formulation described here has been incorporated into the existing propeller design/analysis code XROTOR, which embodies much of the classical propeller theory of Betz [1] and Glauert [2]. The modifications made specifically for the current application were:

- Reformulation to allow arbitrarily large induced velocities relative to the freestream velocity. This is necessary to treat the hovering case where the "freestream" velocity vanishes.

- Incorporation of a shrouded tip into the self-induced velocity formulation. This is necessary to treat the case of a rotor duct.

- Incorporation of external velocities into the velocity triangle definition. This is necessary to represent the counter-rotating rotor system and the presence of a duct.

Nomenclature

B	number of blades per rotor
R	tip radius
V	freestream speed along rotor axis
r, θ, z	cylindrical coordinates, z is along rotor axis
$v_a, v_t(r)$	axial and tangential self-induced velocity components
$w_a, w_t(r)$	axial and tangential externally-induced velocity components
$c(r)$	blade chord
$c_\ell(r)$	blade section lift coefficient
$c_d(r)$	blade section profile drag coefficient
α_o	blade airfoil zero-lift angle
$\beta(r)$	blade angle
$\phi(r)$	net flow angle
$\Gamma(r)$	blade circulation
Ω	rotor rotational speed
λ	advance ratio $V/\Omega R$
φ	self-induced velocity potential
ξ, η, ζ	helical coordinates
δ_{ij}	Kronecker delta, 1 if $i=j$, 0 if $i \neq j$

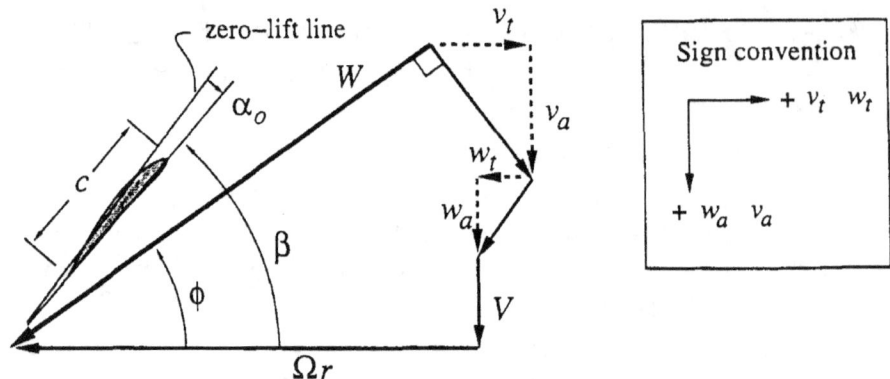

Figure 1: Velocity triangles and blade geometry.

Velocity Decomposition

The rotor blade is represented as a lifting line, with a circulation $\Gamma(r)$ along the radius.

$$\Gamma = \tfrac{1}{2} W c\, c_\ell \tag{1}$$

The local total velocity $W(r)$ relative to the blade is the vector resultant of the freestream velocity V, the rotational component Ωr, the blade's self-induced axial and tangential velocity components v_a and v_t, and any "external" velocities w_a, w_t. The latter are typically due to the a neighboring rotor or a duct. Of couse, a duct can only change w_a but not w_t.

$$W^2 = (V + v_a + w_a)^2 + (\Omega r - v_t - w_t)^2 \tag{2}$$

Figure 1 shows the velocity triangle. The v and w velocity components are all defined to have the same sign convention shown in the figure, but in a counter-rotating rotor system w_t will typically be opposite in sign to v_t as sketched.

Induced-Velocity Solution

The self-induced velocity components are due to the presence of the rotor's own vortex wake, whose flowfield has an associated velocity potential $\varphi(r, \theta, z)$. Its governing equation and boundary conditions are

$$\nabla^2 \varphi \equiv \frac{1}{r}\frac{\partial}{\partial r}\left(r \frac{\partial \varphi}{\partial r}\right) + \frac{1}{r^2}\frac{\partial^2 \varphi}{\partial \theta^2} + \frac{\partial^2 \varphi}{\partial z^2} = 0 \tag{3}$$

$$\Delta \varphi(r) = \Gamma(r) \quad \text{on wake surface} \tag{4}$$

$$\frac{\partial \varphi}{\partial r} = 0 \quad \text{at } r = r_{\text{hub}} \tag{5}$$

$$\text{either} \quad \varphi = 0 \quad \text{at } r \to \infty \quad \text{(free tip case)} \tag{6}$$

$$\text{or} \quad \frac{\partial \varphi}{\partial r} = 0 \quad \text{at } r = R \quad \text{(shrouded tip case)} \tag{7}$$

The wake vortex sheet is assumed to be a helical surface trailing directly from the blade trailing edges with some wake advance ratio λ_w. The pitch between two successive sheets is therefore

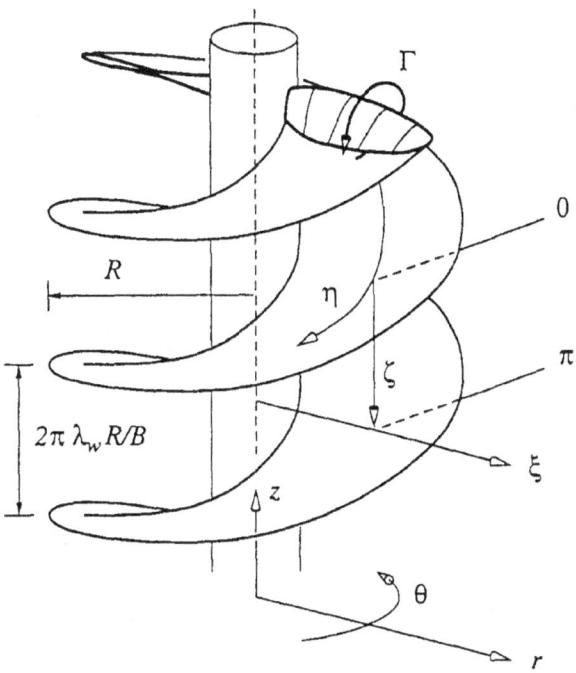

Figure 2: Helical vortex sheet system

$2\pi R\lambda_w/B$. With the convenient helical coordinates

$$\xi = \frac{r}{R} \tag{8}$$

$$\eta = -z - \lambda_w R\theta \tag{9}$$

$$\zeta = \frac{B}{2}\left(\theta - \frac{z}{\lambda_w R}\right) \tag{10}$$

the sheet is located at $\zeta = 0, \pi, 2\pi, 3\pi \ldots$ as shown in Figure 2.

With the assumption of helical invariance ($\partial/\partial\eta = 0$), valid downstream of the rotor, the governing three-dimensional Laplace equation (3) takes the following two-dimensional form.

$$\frac{\partial}{\partial\xi}\left(\xi\frac{\partial\varphi}{\partial\xi}\right) + \frac{B^2}{4}\left(\frac{1}{\xi} + \frac{\xi}{\lambda_w^2}\right)\frac{\partial^2\varphi}{\partial\zeta^2} = 0 \tag{11}$$

This equation is conveniently solved by a mixed spectral–finite difference method. The potential is expressed by a Fourier series in ζ, whose coefficients A_m vary in the radial coordinate ξ.

$$\varphi(\xi,\zeta) = \left(\frac{\zeta}{\pi} - \frac{1}{2}\right)\Gamma(\xi) + \frac{2}{\pi}\sum_{m=2,4\ldots}^{\infty}\frac{1}{m}\left[\Gamma(\xi) + A_m(\xi)\right]\sin(m\zeta) \tag{12}$$

The first term on the righthand side incorporates the potential-jump boundary condition (4) on the wake surface, and preserves the necessary continuity of the wake-normal velocity.

$$\left.\frac{\partial\varphi}{\partial\zeta}\right|_{\zeta=0} = \left.\frac{\partial\varphi}{\partial\zeta}\right|_{\zeta=\pi} \tag{13}$$

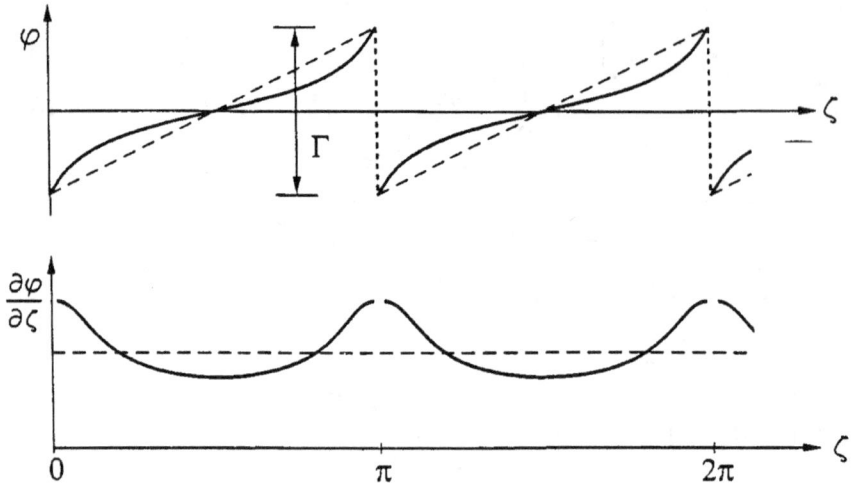

Figure 3: Inter-sheet potential and velocity distributions.

Figure 3 shows the inter-sheet distribution of φ and $\partial\varphi/\partial\zeta$ at some radial location ξ. Substitution of expression (12) into equation (11), and subsequent Fourier analysis

$$\int_0^\pi \Big[\text{equation (11)}\Big]\sin(n\zeta)\,d\zeta \quad ; \quad n = 2,4,6\ldots$$

yields a modified Bessel equation for the Fourier coefficients $A_n(\xi)$.

$$\frac{d}{d\xi}\left(\xi\frac{dA_n}{d\xi}\right) - \frac{n^2 B^2}{4}\left(\frac{1}{\xi} + \frac{\xi}{\lambda_w^2}\right)A_n = \frac{n^2 B^2}{4}\left(\frac{1}{\xi} + \frac{\xi}{\lambda_w^2}\right)\Gamma \qquad (14)$$

Reissner [3], whose formulation was similar but differed in some details, also obtained a modified Bessel equation for the potential harmonics and derived an explicit series solution based on the Green's function approach with modified Bessel functions as the kernel. With modern computers, a simpler approach employed here is to solve equation (14) for each coefficient harmonic $A_n(\xi)$ via the following finite-difference representation.

$$t_i^+ (A_{n_{i+1}} - A_{n_i}) - t_i^- (A_{n_i} - A_{n_{i-1}}) - b_i A_{n_i} = b_i \Gamma_i \quad ; \quad i = 0,1,\ldots K_\infty \qquad (15)$$

where

$$t_i^+ = \frac{1}{2}\frac{\xi_{i+1} + \xi_i}{\xi_{i+1} - \xi_i} \qquad t_i^- = \frac{1}{2}\frac{\xi_i + \xi_{i-1}}{\xi_i - \xi_{i-1}} \qquad (16)$$

and

$$b_i = \frac{n^2 B^2}{4}\left(\frac{1}{\xi_i} + \frac{\xi_i}{\lambda_w^2}\right)\frac{\xi_{i+1} + \xi_{i-1}}{2}. \qquad (17)$$

The boundary conditions (5,6,7), after the appropriate transformation into ξ,ζ coordinates, are discretized as follows.

$$A_{n_1} - A_{n_0} = 0 \qquad (18)$$
$$\text{either} \quad A_{n_{K_\infty}} = 0 \qquad (19)$$
$$\text{or} \quad A_{n_{K+1}} - A_{n_K} = 0 \qquad (20)$$

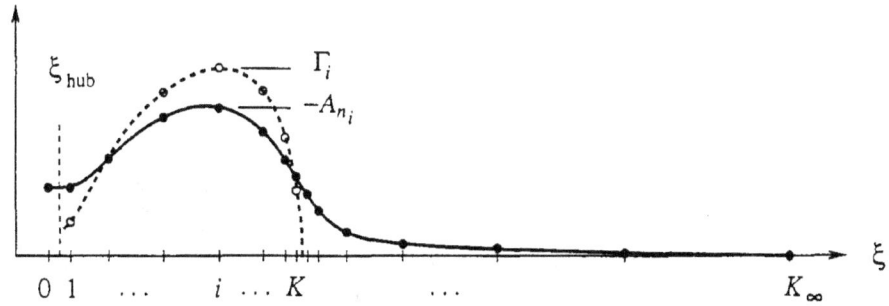

Figure 4: Finite-difference representation of harmonic coefficients A_n. Free-tip case is shown.

The finite-difference grid nodes are arranged so that the hub radius ξ_{hub} is straddled by points $i = 0, 1$ and for the shrouded-tip case the tip radius is straddled by points $i = K, K+1$.

$$\xi_{\text{hub}} = \tfrac{1}{2}(\xi_0 + \xi_1) \tag{21}$$
$$1 = \tfrac{1}{2}(\xi_K + \xi_{K+1}) \tag{22}$$

For the free-tip case, an outer-boundary location of $\xi_{K_\infty} \simeq 3$ is used, which is sufficiently close to "infinity" since the $A_n(\xi)$ functions decay exponentially outside the tip at $\xi = 1$. Figure 4 shows the discrete solution A_{n_i} for a typical case. Because A_{n_i} approaches $-\Gamma_i$ for large n, it is adequate to truncate the infinite sum in (12) to $n = 2, 4, \ldots 2N$, with $N = 32$ being more than sufficient for good accuracy.

The induced velocities at the rotor blade are taken to be half the slipstream velocity in the far wake, since the rotor sees a semi-infinite rather than an infinite wake. The tangential and axial self-induced velocities are then just the derivatives of the velocity potential (12), evaluated on the sheet at $\zeta = 0$ or any multiple of π.

$$v_t(\xi) = \frac{1}{2r}\frac{\partial\varphi}{\partial\theta}\bigg|_{\zeta=0} = \frac{B}{4r}\frac{\partial\varphi}{\partial\zeta}\bigg|_{\zeta=0} = \frac{B}{4\pi r}\left\{\Gamma(\xi) + 2\sum_n\left[\Gamma(\xi) + A_n(\xi)\right]\right\} \tag{23}$$

$$v_a(\xi) = \frac{1}{2}\frac{\partial\varphi}{\partial z}\bigg|_{\zeta=0} = \frac{B}{4}\frac{1}{\lambda_w R}\frac{\partial\varphi}{\partial\zeta}\bigg|_{\zeta=0} = \frac{\xi}{\lambda_w} v_t(\xi) \tag{24}$$

It is useful to note that the vector resultant of v_t and v_a is perfectly perpendicular to the sheet at any radius ξ.

To obtain a general concise expression for the nodal velocities v_{t_i}, v_{a_i}, the finite-difference equation (15) is solved a priori for each n and for each unit nodal circulation $\Gamma_j = 1$

$$t_i^+ \left(\tilde{A}_{n_{i+1\,j}} - \tilde{A}_{n_{ij}}\right) - t_i^- \left(\tilde{A}_{n_{ij}} - \tilde{A}_{n_{i-1\,j}}\right) - b_i \tilde{A}_{n_i} = b_i\, \delta_{ij} \tag{25}$$

so that the general solution for any Γ_i distribution is

$$A_{n_i} = \sum_{j=1}^{K} \tilde{A}_{n_{ij}}\, \Gamma_j \tag{26}$$

with j being summed only over the blade stations where Γ is defined. This is then substituted into the induced-velocity relation (23) to give similar convenient expressions for the nodal induced velocities.

$$v_{t_i} = \sum_{j=1}^{K} a_{ij} \Gamma_j \qquad v_{a_i} = \frac{\xi_i}{\lambda_w} \sum_{j=1}^{K} a_{ij} \Gamma_j \qquad (27)$$

$$\text{where} \qquad a_{ij} = \frac{B}{4\pi \xi_i} \left\{ \delta_{ij} + 2 \sum_n \left[\delta_{ij} + \tilde{A}_{n_{ij}} \right] \right\} \qquad (28)$$

In the present semi-free wake formulation, λ_w is determined by the disk loading and so depends on v_t and v_a themselves. To enable the construction of an implicit Newton system for the overall flow solution, the following sensitivities with respect to λ_w must be calculated. Using the convenient notation $\widehat{(\ )} = \partial/\partial\lambda_w$, these follow directly from (23,24).

$$\frac{\partial v_{t_i}}{\partial \lambda_w} = \frac{B}{4\pi r_i} 2 \sum_n \widehat{A}_{n_i} \qquad \frac{\partial v_{a_i}}{\partial \lambda_w} = \frac{\xi_i}{\lambda_w} \frac{\partial v_{t_i}}{\partial \lambda_w} - \frac{\xi_i}{\lambda_w^2} v_{t_i} \qquad (29)$$

The derivative array $\widehat{A}_i$ is calculated from equation (15) after it is implicitly differentiated with respect to λ_w.

$$t_i^+ \left(\widehat{A}_{n_{i+1}} - \widehat{A}_{n_i} \right) - t_i^- \left(\widehat{A}_{n_i} - \widehat{A}_{n_{i-1}} \right) - b_i \widehat{A}_{n_i} = \left[\frac{n^2 B^2}{4} \left(-\frac{2\xi_i}{\lambda_w^3} \right) \frac{\xi_{i+1} + \xi_{i-1}}{2} \right] (\Gamma_i - A_{n_i}) \qquad (30)$$

This is the same as the original equation (15) but with a different righthand side.

The potential flowfield generated by the present finite-difference/spectral solution method perfectly matches the special cases of Goldstein [4], who obtained an analytic solution for the flow about a non-deforming helical surface. Consistency with this special test case confirms that the implementation of the present solution method is correct.

Blade-Section Relations

As with any lifting-line representation, a locally-2D section lift relation is assumed.

$$c_\ell = m_{2D} (\beta - \alpha_o - \phi) \qquad (31)$$

The local blade angle β and zero-lift angle α_o are known, and the net flow angle ϕ follows from the all the velocity components in the velocity triangle in Figure 1.

$$\phi = \arctan\left(\frac{V + v_a + w_a}{\Omega r - v_t - w_t} \right) \qquad (32)$$

Inserting these expressions into the Kutta-Joukowsky relation (1) evaluated at the nodal locations ξ_i gives the following node residual.

$$\mathcal{R}_i(\Gamma_j; \lambda_w) \equiv \Gamma_i - \frac{1}{2} \left[(V + v_{a_i} + w_{a_i})^2 + (\Omega r_i + v_{t_i} + w_{t_i})^2 \right]^{1/2}$$
$$\times c \, m_{2D} \left[\beta_i - \alpha_o - \arctan\left(\frac{V + v_{a_i} + w_{a_i}}{\Omega r_i + v_{t_i} + w_{t_i}} \right) \right] = 0 \qquad (33)$$

The dependency on all nodal circulation values Γ_j and on the wake advance ratio λ_w comes from the $v_{a_i}(\Gamma_j; \lambda_w)$ and $v_{t_i}(\Gamma_j; \lambda_w)$ expressions (27,29).

The overall thrust and power are calculated by integrating the axial and tangential section lift and drag force components along the blade.

$$T = T_{\text{inv}} - T_{\text{vis}} \tag{34}$$

$$P = P_{\text{inv}} + P_{\text{vis}} \tag{35}$$

$$T_{\text{inv}} = B \int_0^R \tfrac{1}{2}\rho W^2 \, c \, c_\ell \cos\phi \, dr \tag{36}$$

$$T_{\text{vis}} = B \int_0^R \tfrac{1}{2}\rho W^2 \, c \, c_d \sin\phi \, dr \tag{37}$$

$$P_{\text{inv}} = B \int_0^R \tfrac{1}{2}\rho W^2 \, c \, c_\ell \sin\phi \, \Omega r \, dr \tag{38}$$

$$P_{\text{vis}} = B \int_0^R \tfrac{1}{2}\rho W^2 \, c \, c_d \cos\phi \, \Omega r \, dr \tag{39}$$

The wake advance ratio λ_w, which determines the wake pitch, is defined by

$$\lambda_w = \lambda \frac{P_{\text{inv}}}{V \, T_{\text{inv}}} = \frac{P_{\text{inv}}}{\Omega R \, T_{\text{inv}}} \tag{40}$$

which reduces to the actuator-disk limit for many blades, uniform disk loading, and small advance ratio λ. For the case of a minimum induced loss rotor, this definition also results in the trailing vortex sheet being aligned with the local velocity W. For more general rotors the wake's self-induction does not preserve its shape, but equation (40) is nevertheless a rational means to estimate the wake's self-induction effect on its average pitch in some sense.

In the approximation hierarchy, the present method sits midway between a full free-wake model complete with wake rollup treatment, and the classical light-loading methods such as those of Betz [1] and Larrabee [5] which simply assume $\lambda_w = \lambda$. Treating wake rollup is deemed excessive given the circumferential-averaging approximation necessary to model the counter-rotating case, while the classical methods are inadequate for a hovering rotor, where $\lambda = 0$ by definition but λ_w remains finite. The present approach is deemed to strike an appropriate middle ground.

Analysis Solution

A rotor of a given geometry $\beta(r)$, $c(r)$ is analyzed by simultaneously solving the following system of residuals for Γ_i and λ_w

$$\mathcal{R}_i(\Gamma_j; \lambda_w) = 0 \tag{41}$$

$$\mathcal{Q}(\Gamma_j; \lambda_w) = 0 \tag{42}$$

where the auxilliary residual $\mathcal{Q}$ is simply equation (40) put in residual form.

$$\mathcal{Q}(\Gamma_j; \lambda_w) \equiv \lambda_w - \frac{P_{\text{inv}}}{\Omega R \, T_{\text{inv}}} \tag{43}$$

The nonlinear system (42) is solved by a standard Newton method, with the induced velocity coefficient matrix a_{ij} begin recomputed at every Newton iteration. Three Newton iterations typically suffice to achieve machine zero. The resulting Γ_i and λ_w are then used to compute the thrust and power using equations (34,35).

Design Solution

In a design calculation, the c_ℓ is typically prescribed, and the objective is to determine Γ and hence the blade geometry via the section relations.

$$c = \frac{2\Gamma}{W c_\ell} \qquad (44)$$

$$\beta = \frac{c_\ell}{m_{2D}} + \alpha_o + \phi \qquad (45)$$

Since these relations already embody the residual $\mathcal{R}$ in the implicit system (42), it must be replaced by some other residual $\mathcal{R}'$. A suitable choice is the condition that the *locally-defined* wake advance ratio be equal to the average value λ_w,

$$\mathcal{R}'_i(\Gamma_j; \lambda_w) \equiv \lambda_w - \left[\frac{V + v_a + w_a}{\Omega r - v_t - w_t}\right]_i \qquad (46)$$

which simply states that the vortex sheet trails from the trailing edge along the total resultant velocity W at every radial location. Likewise, the residual $\mathcal{Q}$ is replaced by $\mathcal{Q}'$, which is some constraint on the total thrust or power to take on a specified value T_{spec} or P_{spec}.

$$\mathcal{Q}'(\Gamma; \lambda_w) \equiv T - T_{\text{spec}} \qquad (47)$$

The particular definition of $\mathcal{R}'$ above will result in a minimum-induced loss rotor analogous to that of Betz [1] for the case of light disk loadings. The key difference is that the present formulation produces a minimum-induced loss rotor for the heavily-loaded hovering case, while the Betz result breaks down in this limit.

Counter-rotating Rotor Representation

The presence of a duct or another rotor ahead or behind the one being analyzed is represented by nonzero "external" induced velocities w_a and w_t. For the duct, these correspond to the flowfield associated with the bound vorticity representing the duct, and can be determined via a panel or vortex solution, or via some other estimate. The contribution to w_a and w_t from the neigboring rotor is less well-defined, since this is unsteady in the actual case, and must be defined in a time-averaged sense.

A rational approach is to assume that w_a and w_t are the circumferentially-averaged induced velocities of the other rotor's bound and trailing vortices. The velocity components far downstream are the derivatives of the velocity potential (12), which when averaged in ζ give the following.

$$w_{t_\infty}(\xi) = \frac{1}{\pi} \int_0^\pi \frac{B}{2r} \frac{\partial \varphi}{\partial \zeta} d\zeta = \frac{B\Gamma(\xi)}{2\pi r} \qquad (48)$$

$$w_{a_\infty}(\xi) = \frac{1}{\pi} \int_0^\pi \frac{B}{2} \frac{1}{\lambda_w R} \frac{\partial \varphi}{\partial \zeta} d\zeta = \frac{\xi}{\lambda_w} \frac{B\Gamma(\xi)}{2\pi r} \qquad (49)$$

These correspond to the avarage $\partial \varphi / \partial \zeta$ value shown dashed in Figure 3.

The w_{a_∞} expression (49) corresponds to a location far downstream which sees the entire infinite wake. A point closer to the rotor will see some fraction of the wake and w_a will be correspondingly smaller. The effect is represented here by a reduction factor f_a which multiplies the maximum value. The tangential velocity component is also influenced by the bound vorticity, so that the overall w_t either the full value given by (48) or zero, depending if the location is downstream or upstream of the rotor. This can be likewise represented by a factor f_t.

$$w_t(\xi) = \frac{f_t}{\xi} \frac{B\Gamma(\xi)}{2\pi R} \qquad (50)$$

$$w_a(\xi) = \frac{f_a}{\lambda_w} \frac{B\Gamma(\xi)}{2\pi R} \qquad (51)$$

In a dual-rotor system, the axial factor takes on the values $0 < f_a < 1$. The tangential factor is $f_t = 0$ or $f_t = \pm 1$, depending on where the "other" rotor is relative to the one being analyzed. If the rotors are close together in the axial direction, then $f_a = 0.5$ for both. If the "other" rotor is downstream, then the analyzed upstream rotor will not see any tangential velocity and hence $f_t = 0$. If the "other" rotor is upstream, then the analyzed upstream rotor will see the full tangential velocity, and so $f_t = -1$ for a counter-rotating system and $f_t = +1$ for a co-rotating system. As is well-known, the counter-rotating case allows a near-cancellation between w_t and v_t in the wake, as sketched in Figure 1, and thus gives a possible efficiency increase compared to a single or a co-rotating rotor.

Counter-rotating Rotor Design/Analysis

The actual procedure used to analyze or design a counter-rotating rotor system involves simple iterating between the two rotors, with the solution from one providing the externally-imposed induced velocity distributions $w_a(\xi)$, $w_t(\xi)$ for the other via equations (50,51). The top rotor couples less strongly to the bottom rotor than vice versa, so designing or analyzing the top rotor first is more effective. The convergence is very rapid in any case, requiring only a few iterations. A similar approach is taken for the presence of the duct, with the iteration now being between the rotor analysis and the duct analysis.

References

[1] A. Betz. Airscrews with minimum energy loss. Report, Kaiser Wilhelm Institute for Flow Research, 1919.

[2] H. Glauert. *Elements of Airfoil and Airscrew Theory*. Cambridge University Press, Cambridge, 1937.

[3] H Reissner. Theory of propellers. Brown University Report, 1942.

[4] S. Goldstein. On the vortex theory of screw propellers. *Proceedings of the Royal Society*, 123, 1929.

[5] E.E. Larrabee and S.E. French. Minimum induced loss windmills and propellers. *Journal of Wind Engineering and Industrial Aerodynamics*, 15:317–327, 1983.

Propeller
Airfoil Coordinates

The Mapping format

As depicted in the figure below, the coordinates listed for "exotic" props start at the upper-surface trailing edge at x =1, y = 0 (1,0), go around the top of the airfoil to the leading edge (0,0) and back to the trailing edge (1,0). Airfoil coordinates through the leading-edge region may not include the x=0, y=0 (0,0) point.

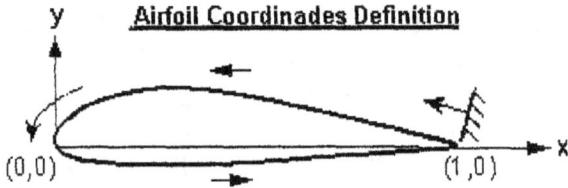

Some Definitions

The variables used on the forms that define the velocity distribution from which the airfoil shape is derived. As depicted in the figure below, the airfoil is divided into a given number of segments, which are numbered in counter-clockwise order. The arc limits PHIL, PHI2, PHI3 and PHI4 on the circle are mapped to the points S1, S2, S3 and S4 on the airfoil. PHI is measured from the point at PHI4. In the method, the minimum number of segments is four - two on the upper surface of the airfoil and two on the lower surface as shown. The circle arc limit PHI begins at PHI = 0 and ends at PHI = 60 (rather than 360 deg); thus, the circle is divided into 6 deg sectors.

Associated with each segment is a corresponding "design angle of attack," which is best described by way of example. Suppose that the design angle of attack for the second segment ALPHA2 is set at 10 deg. When the resulting airfoil is then operated at 10 deg. the velocity distribution along the second segment will be *constant*. For the third segment, the design angle of attack can be set to 0 deg. which would give a constant velocity along that segment of 0 deg.

It should be noted that the design angle of attack is referenced to the airfoil zero-lift line (rather than the airfoil chord line). Thus, specifying the design angle of attack allows one to prescribed a constant velocity distribution on a segment for a given lift coefficient, since the lift coefficient of the airfoil is approximately 2 pi ALPHA.

Misc comments:
The freestream is set to unity, i.e., Vinfinity = 1;
The velocity distributions presented along with the airfoil are for angles of attack: 2, 4, 6, 8, 10, 12, and 14 deg.

More details:
. Selig, M.S. and Maughmer, M.D., "Multipoint Inverse Airfoil Design Method Based on Conformal Mapping.

http://amber.aae.uiuc.edu/~m-selig/profoil/055-coordsFmt.html

"EPPLER - 193 Cordinates - 10.22%   CM0=-0.0781

Sta.	X	Y	Sta.	X	Y
1	100.00	0.000	31	0.465	-0.917
2	99.661	0.051	32	0.026	-0.192
3	98.674	0.221	33	0.129	-0.373
4	97.108	0.522	34	0.819	-0.837
5	95.023	0.932	35	2.044	-1.250
6	92.452	1.415	36	3.791	-1.586
7	89.414	1.957	37	6.049	-1.839
8	85.945	2.558	38	8.801	-2.008
9	82.096	3.215	39	12.026	-2.097
10	77.923	3.914	40	15.697	-2.110
11	73.484	4.643	42	19.778	-2.059
12	68.839	5.382	42	24.227	-1.954
13	64.052	6.112	43	28.998	-1.806
14	59.187	6.808	44	34.035	-1.627
15	54.306	7.437	45	39.280	-1.429
16	49.458	7.955	46	44.672	-1.223
17	44.673	8.333	47	50.145	-1.018
18	39.979	8.552	48	55.630	-0.823
19	35.403	8.604	49	61.059	-0.644
20	30.968	8.488	50	66.364	-0.485
21	26.696	8.214	51	71.479	-0.350
22	22.620	7.807	52	76.339	-0.239
23	18.781	7.285	53	80.882	-0.153
24	15.218	6.664	54	85.050	-0.090
25	11.968	5.958	55	88.788	-0.048
26	9.061	5.183	56	92.048	-0.018
27	6.525	4.354	57	94.794	0.010
28	4.383	3.489	58	97.003	0.033
29	2.652	2.610	59	98.640	0.034
30	1.344	1.741	60	99.655	0.014
				100.000	0.000

BETA= 3.39 DEG

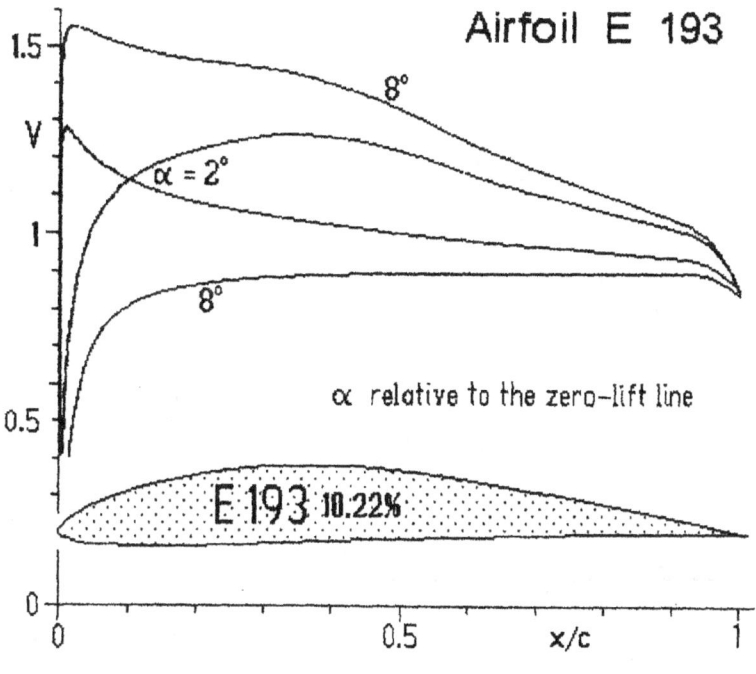

Airfoil E 193

α relative to the zero-lift line

E 193 10.22%

Designed for Re ≥ 200,000
For Re = 100,000 only, the
C_l - values near C_l = 1
provide good C_d.
A tubulator with thickness
of 0.5 % chord at 30%
chord could possibly
improve the airfoil at
Re = 100,000.

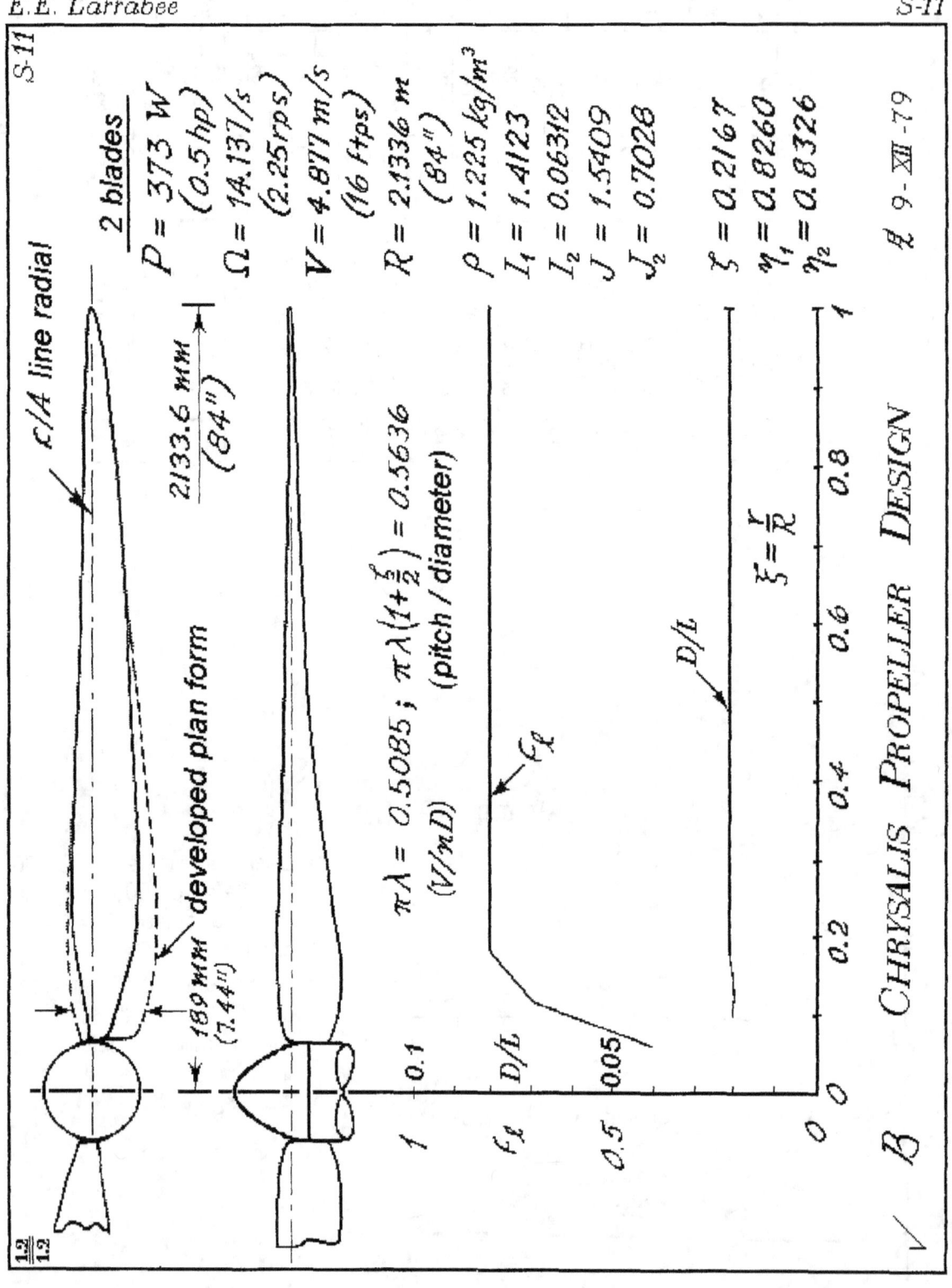

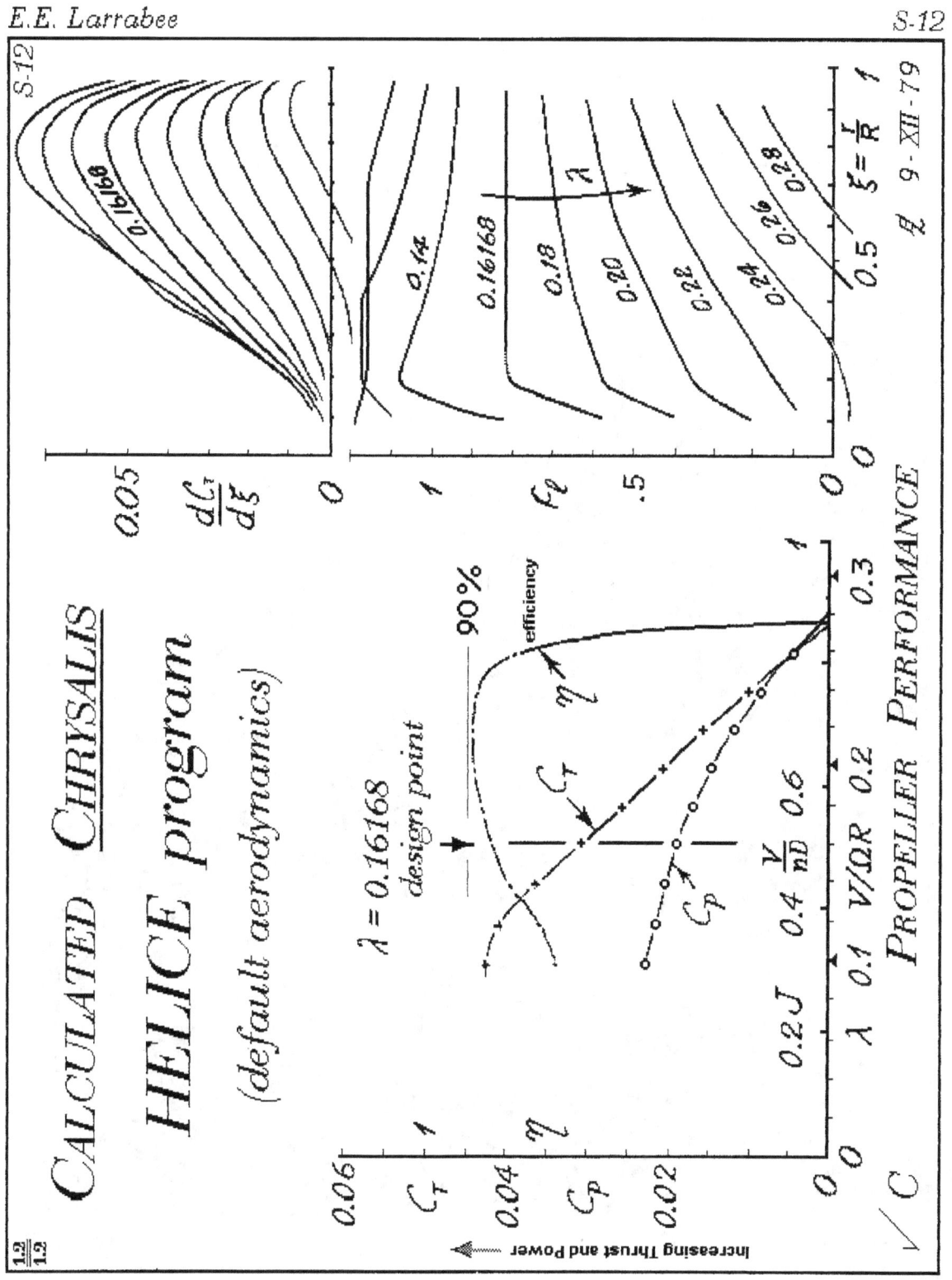

MUSCLE POWERED BLIMPS

SCIENTIFIC AMERICAN

HUMAN-POWER LAND VEHICLES

$2.50

December 1983

The Aerodynamics of Human-powered Land Vehicles

A bicycle and its rider are strongly impeded by their resistance to the flow of air. Aerodynamic stratagems have brought vehicles that can go 60 miles per hour on a level road without assistance

by Albert C. Gross, Chester R. Kyle and Douglas J. Malewicki

For decades the principles of aerodynamics have been applied with great success to improving the speed and efficiency of aircraft, automobiles, motorcycles and even competitive skiers and skaters. Vehicles powered by human energy, however, were virtually ignored until quite recently, which is strange in view of the fact that air resistance is by far the major retarding force affecting them. With a bicycle, for example, it accounts for more than 80 percent of the total force acting to slow the vehicle at speeds higher than 18 miles per hour. Here we undertake to explain this neglect and to show what attention to aerodynamics is beginning to do for the performance of human-powered land vehicles.

Looking first at the bicycle, one sees that it has remained almost the same in form for nearly a century. The Rover Safety Cycle, which was introduced in England in 1884, could easily pass for a modern bicycle; it lacks only a seat brace, which would have formed the modern diamond frame, and a few components such as brakes and multiple gears. Almost from the beginning the designers and users of bicycles recognized the importance of aerodynamics, but artificial constraints on design largely prevented the application of the necessary technology. It was as obvious then as it is now that wind forces at the bicycle-racing speed of from 20 to 30 m.p.h. are enormous.

Before 1900 the crouched posture of the bicycle racer had become common as a means of reducing air resistance. Another practice adopted before 1900 was to put a multiple-rider bicycle ahead of a single racer to shield him from the wind. In 1895 the Welsh wheelman Jimmy Michael rode 28.6 miles in one hour behind a four-man lead bicycle. In 1899 Charles ("Mile-a-Minute") Murphy of the U.S. gained international fame by pedaling one mile at 63.24 m.p.h. on a bicycle traveling behind a train of the Long Island Rail Road on a board path built for the occasion.

In 1912 Étienne Bunau-Varilla of France patented a streamlined enclosure for a bicycle and its rider that was inspired by the shape of the first dirigible balloons. Versions of this bicycle and its descendants set speed records in Europe from 1912 to 1933. In 1933 Marcel Berthet of France covered 31.06 miles in one hour riding a streamlined rig named the Vélodyne; his pace was more than 3 m.p.h. faster than anyone riding a standard bicycle had gone for one hour.

In the same year the French inventor Charles Mochet built a supine recumbent bicycle (with the rider pedaling while lying on his back) that he later streamlined. With a professional racer, François Faure, this "Vélocar" set a number of speed records between 1933 and 1938. Mochet and Faure hoped the records would be recognized by the Union Cycliste Internationale, the world governing body for bicycle racing. They were not.

Indeed, in 1938 the Union banned the use of aerodynamic devices and recumbent bicycles in racing; the rule is still in force. The ban has been a serious deterrent to the development of high-speed bicycles and is one of two major reasons the bicycle has remained nearly unchanged for so long. (The other reason is that in the developed countries the shift to the automobile has made the bicycle less important for transportation than it once was.)

By its ruling the Union essentially classified improvements in the aerodynamics of bicycles and other technological changes as "cheating." (It is perhaps fortunate that the Union was not active when a Scotch-Irish veterinary surgeon, John Boyd Dunlop, developed the pneumatic tire for bicycles in 1887, otherwise people might now be riding bicycles and possibly automobiles with solid steel wheels.) To its credit, however, the Union has gradually begun to relax its restrictions on changes in aerodynamics, although recumbents are still forbidden. Since 1976 skintight one-piece suits have become common in international bicycle racing. Streamlined helmets, teardrop cross sections for frame tubing, streamlined brake levers and other aerodynamically improved components have been allowed. In fact, technological change in all forms of human-powered vehicle is flourishing at a rate unmatched since the heyday of the bicycle in the 19th century.

This rapid change can be partly attributed to a series of events in California. In 1973 one of us (Kyle) and Jack H. Lambie, a consultant in aerodynamics who was working independently, built and tested the first two streamlined bicycles in the U.S. Unlike their predecessors, Kyle and Lambie actually measured the reduction in drag achieved by streamlining. They did so by conducting numerous coast-down tests, in which an unpowered vehicle is allowed to decelerate on a level surface. In this condition the deceleration of the vehicle is proportional to the total retarding forces acting on it; instruments measure either the speed or the deceleration. Kyle and Lambie, publishing their results independently, both concluded that the total drag forces on a bicycle could be reduced by more than 60 percent with a vertical, wing-shaped fairing that completely encloses the bicycle and the rider. (It was not until some two years later that either Kyle or Lambie learned that similar vehicles had been built earlier in Europe.)

In 1974 Ronald P. Skarin, an Olympic cyclist for the U.S., set five world speed records riding the Kyle streamlined bicycle at the Los Alamitos Naval Air Station. Because of this success, Kyle and

MUSCLE POWERED BLIMPS

Lambie decided to organize a race for unrestricted human-powered vehicles. On April 5, 1975, at Irwindale, Calif., 14 distinctive vehicles competed in this historic first race. Many of them were recumbents, some with the rider pedaling supine (face up) and some with the rider prone (face down). Some were propelled by both hand and foot power. The winner at 44.87 m.p.h. was a streamlined tandem bicycle designed by Philip Norton, a high school teacher in Edgewood, Calif. The pedalers were Norton and Christopher Deaton, who is a skilled racing cyclist but not a world-class competitor. (The fastest an unaided standard racing bicycle has been ridden is 43.45 m.p.h., a record set in 1982 by Sergei Kopylov of the U.S.S.R., a cyclist of world class.)

Faced with the policy of the Union Cycliste Internationale against streamlining, the competitors in this race founded the International Human Powered Vehicle Association in 1976. Its purpose was to sanction competitions in which human-powered vehicles would be under no restrictions of design. Since then in dozens of races held in many countries the machines have become much more sophisticated and speeds have risen steadily. Four vehicles have broken the U.S. automobile speed limit of 55 m.p.h. (Each one received an honorary speeding ticket from the California Highway Patrol.) Among them is a third-generation streamlined quadricycle designed by Norton.

At present the world's fastest human-powered vehicle is the Vector Tandem, a gracefully streamlined two-person recumbent. It was built by a team headed by Allan A. Voigt, an engineer who as president of Versatron Research, Inc., primarily designs aerospace servomotors. (The pedalers ride supine and facing in opposite directions.) In 1980, with a flying start of about one mile of acceleration, it covered 200 meters along the track of the Ontario Motor Speedway in California at 62.92 m.p.h. Later that year the Vector Tandem averaged 50.5 m.p.h. for 40 miles on Interstate Route 5 between Stockton and Sacramento.

These extraordinary speeds are almost entirely the result of attention to aerodynamics. A cyclist traveling at 20 m.p.h. typically displaces approximately 1,000 pounds of air per minute. When the machine and the rider are not streamlined, they leave a substantial wake and exact a high cost in human energy.

Two types of aerodynamic drag affect the performance of a bicycle: pressure (or form) drag and skin-friction drag. Pressure drag results when the flow of air fails to follow the contours of the moving body. The separation changes the distribution of the air pressure on the body. If the separation takes place toward the rear of the body, the air pressure there becomes lower than it is on the forward surface, causing drag.

STREAMLINED RACING BICYCLE designed by one of the authors (Kyle) and ridden by Ronald P. Skarin, an Olympic cyclist for the U.S., is shown setting the world record of 31.88 miles per hour for one hour of pedaling from a standing start. The key to the performance was the streamlined fairing that reduced the aerodynamic resistance of the rider and the bicycle. Skarin established the new speed record in 1979 at the Ontario Motor Speedway in Ontario, Calif. Except for the fairing the vehicle was basically a standard racing bicycle.

Skin-friction drag results from the viscosity of the air. It is caused by the shearing forces generated in the boundary layer: the layer of air immediately next to the surface of the body.

Blunt configurations such as the cylinders, spheres and other shapes found on a bicycle are aerodynamically inefficient because the airflow separates from the surfaces. Low-pressure regions form behind the objects, resulting in pressure drag hundreds of times greater than skin-friction drag. In contrast, air flows smoothly around a streamlined shape. The air closes in behind as the body passes. Pressure drag is greatly reduced and skin-friction drag becomes more important.

For the highest efficiency a vehicle should be designed to minimize the transfer of unrecoverable energy to the air by the two types of drag. At the present level of technology aerodynamic drag absorbs from 40 to 50 percent of the fuel energy consumed by an automobile or a truck at 55 m.p.h. Since the bicycle has lower power, weight and

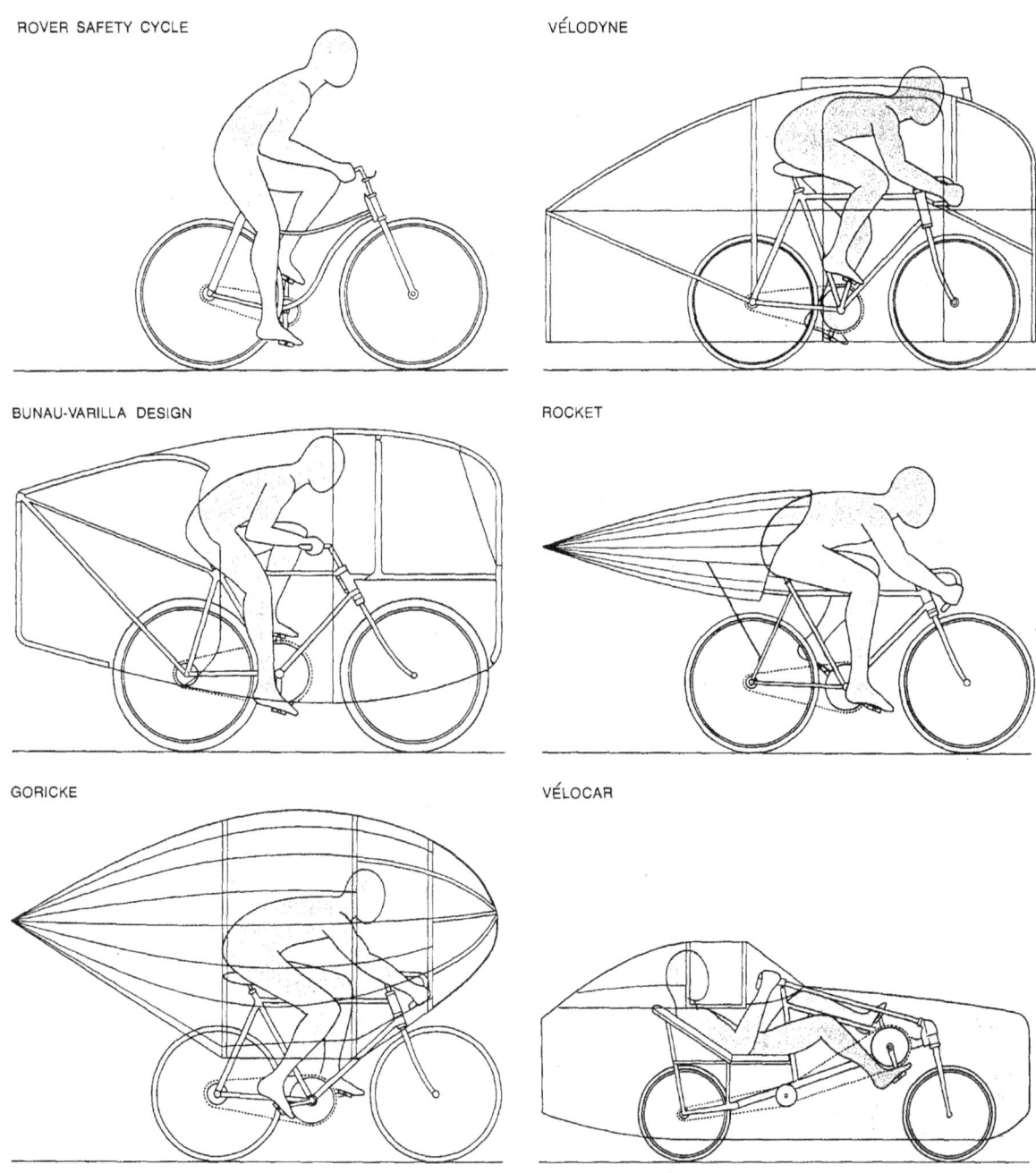

EARLY IMPROVEMENTS in human-powered land vehicles resulted in the introduction of the Rover Safety Cycle in England in 1884. In 1912 and 1913 Étienne Bunau-Varilla of France obtained patents for a streamlined design; similar bicycles set many speed records. The Goricke was developed in Germany in 1914. The Vélodyne was ridden 31.06 miles in one hour (a new record) by Marcel Berthet of France in 1933. From the same year is the Rocket, designed by Oscar Egg. Another French vehicle, the Vélocar, set several speed records between 1933 and 1938. Most of the drawings are based on data from the Wolfgang Gronen Archive at Binningen in Switzerland.

rolling resistance and poor streamlining, aerodynamic drag accounts for an even higher percentage of the energy consumed at speeds above 10 m.p.h.

A term employed to describe the aerodynamic efficiency of a shape is the drag coefficient. An inefficient shape such as a sphere will have a drag coefficient of, say, 1.3, whereas a streamlined shape such as a teardrop will have one of less than .1. Hence an object of teardrop shape can move with less than a tenth of the loss of energy incurred by an object of cylindrical shape.

For land-transportation vehicles the aerodynamic resistance is almost directly proportional to the product of the frontal area and the drag coefficient. For convenience we call the product the effective frontal area. In discussing which of two vehicles has less aerodynamic drag it is not sufficient to compare drag coefficients; the size of the vehicle must also be taken into account. That is done in the concept of the effective frontal area. An ordinary bicycle and its rider will have an effective frontal area of from 3.4 to six square feet, whereas a streamlined human-powered vehicle can have one of less than .5 square foot.

The force of aerodynamic drag increases as the square of the velocity. Power is proportional to the product of drag force and velocity, so that the power necessary to drive an object through the air increases as the cube of the velocity. Hence a modest increase in speed requires an enormous increase in power. A cyclist who suddenly doubles his output of power when he is traveling at 20 m.p.h. will increase his speed to only about 26 m.p.h.

Conversely, reductions in aerodynamic drag affect speed less than one might think. If the air drag is cut in half at 20 m.p.h., a cyclist who does not change his power output will speed up to about 24.4 m.p.h. The reason is that the rolling resistance remains constant. If that resistance could be ignored, doubling the horsepower or reducing the effective frontal area by half would again get the speed up to about 26 m.p.h.

In sum, high speeds require extremely high aerodynamic efficiency. The Vector Tandem, receiving an input of slightly more than one horsepower from each of its two riders, attained a speed of 62.92 m.p.h. For a standard bicycle to achieve that speed would require more than 6 h.p. That level of power from a human rider is clearly impossible.

Designers and riders can reduce the aerodynamic drag on human-powered vehicles in three major ways. First, they can cut down the amount of energy wasted by the vehicle's interaction with the air. They do it by streamlining (reshaping the front and rear of blunt objects to minimize the pressure drag) and by smoothing rough surfaces to minimize skin-friction drag. Second, the amount of air encountered during each second of forward travel can be reduced. This is done by lowering the effective frontal area of the vehicle-rider combination. The same effect can be achieved by riding at higher altitudes. Third, the rider can find air moving in such a way that it provides a tail wind. Here the most effective approach is drafting, that is, riding closely in the wake of another vehicle.

At high altitudes the atmosphere is less dense and bicyclists encounter less air. In Mexico City (elevation 7,414 feet, where the air is only 80 percent as dense as it is at sea level) cycling records are from 3 to 5 percent faster than records made at lower altitudes. At La Paz, Bolivia (elevation 12,000 feet), sea-level

ZZIPPER

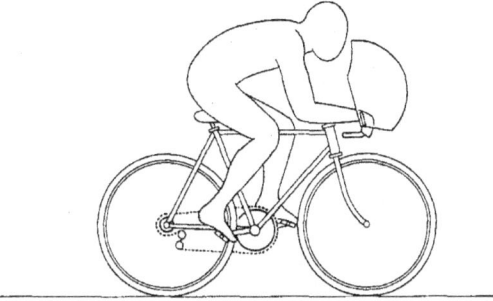

VECTOR SINGLE

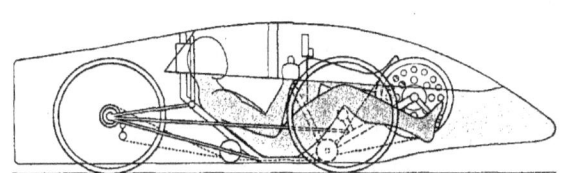

KYLE STREAMLINER

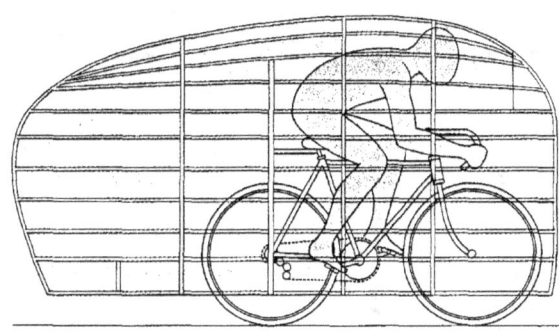

EASY RACER

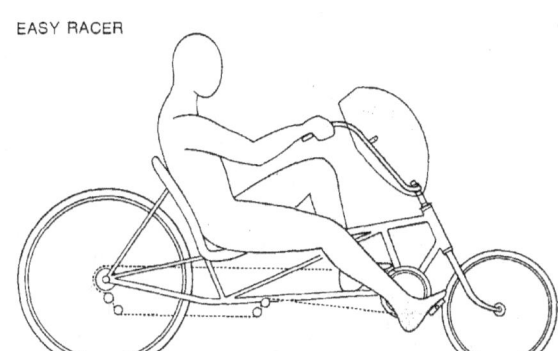

AVATAR 2000

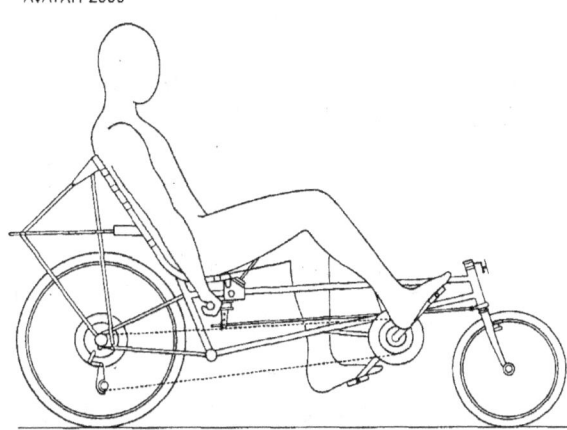

SCHÖNDORF ALL-WEATHER VEHICLE

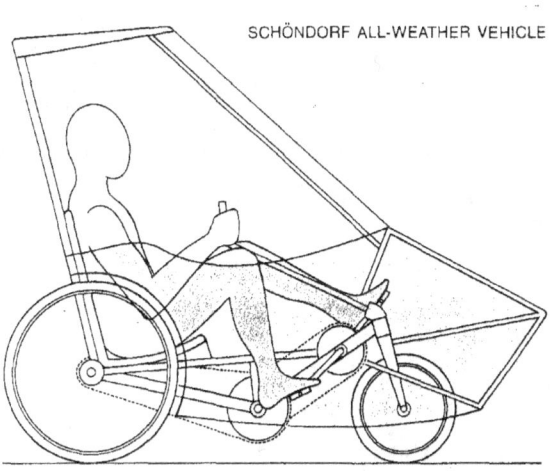

MODERN HUMAN-POWERED VEHICLES make intensive use of streamlining to reduce the aerodynamic drag of the vehicle-rider combination. The simplest is the Zzipper, which is a partial fairing mounted in front of the rider. The Kyle Streamliner dates from 1973. A design that is meant for touring and commuting rather than for racing is the Avatar 2000; it utilizes the advantages of a recumbent position for the rider. The Vector Single, which has a full fairing, is theoretically capable of reaching almost 62 m.p.h. with an input of one horsepower from the rider. The Easy Racer is a recumbent designed mainly for touring or commuting, but it has also been raced. The last vehicle is one of the all-weather recumbents designed by Paul Schöndorf in Germany for elderly and handicapped people.

records could theoretically be improved by 14 percent. On the moon, where there is no atmosphere and only one-sixth the gravitational attraction, a suitably equipped bicyclist could theoretically ride at 238 m.p.h. with a very modest input of .1 h.p.

Analyzing the relation in which 80 percent of the power generated by a cyclist traveling on level ground at 18 m.p.h. goes to overcome air resistance, one finds that about 70 percent of the power consumption is due to the air's resistance to the rider and 30 percent to the air's resistance to the bicycle. This finding leads to the conclusion that to improve the performance of the standard bicycle one must first improve the aerodynamics of the rider.

For riders who race, the restrictions of the Union Cycliste Internationale leave little room for improvements beyond what has already been done in adopting the crouched position, the streamlined helmet, the skintight suit and the streamlining of components of the bicycle. As Voigt has calculated, even with a "perfect" bicycle (no aerodynamic drag on the machine at any speed and tires with no rolling resistance) the aerodynamic drag on the rider alone would severely hamper improvements in performance. According to Voigt, a crouched rider on a conventional racing bicycle could reach a maximum velocity of about 34 m.p.h. with a power input of 1 h.p. On a perfect bicycle the same rider making the same effort could achieve 38 m.p.h.

For the millions of noncompetitive cyclists who simply want a more efficient ride, several aerodynamic improvements are possible. They can be ranked in order of cost, beginning with the cheapest: a partial fairing such as the Zzipper, developed and manufactured by Glen Brown of Santa Cruz, Calif. It is a small, transparent, streamlined shield mounted in front of the rider. For about $60 a rider can lower the aerodynamic drag by about 20 percent, achieving a speed increase of some 2.5 m.p.h. for a 1-h.p. input.

Another effective way of reducing aerodynamic drag is to ride a recumbent bicycle. (The machine would cost several hundred dollars more than a basic touring bicycle.) The pioneers in this field are Gardner Martin of Freedom, Calif., designer of the Easy Racer, and David Gordon Wilson of the Massachusetts Institute of Technology, designer of the Avatar 2000. Because of the smaller frontal area presented by the recumbent rider, wind resistance decreases by 15 to 20 percent, resulting in about the same speed increase as is achieved by the Zzipper fairing.

The recumbent bicycle offers other advantages. It is more comfortable to ride than a standard bicycle. In acci-

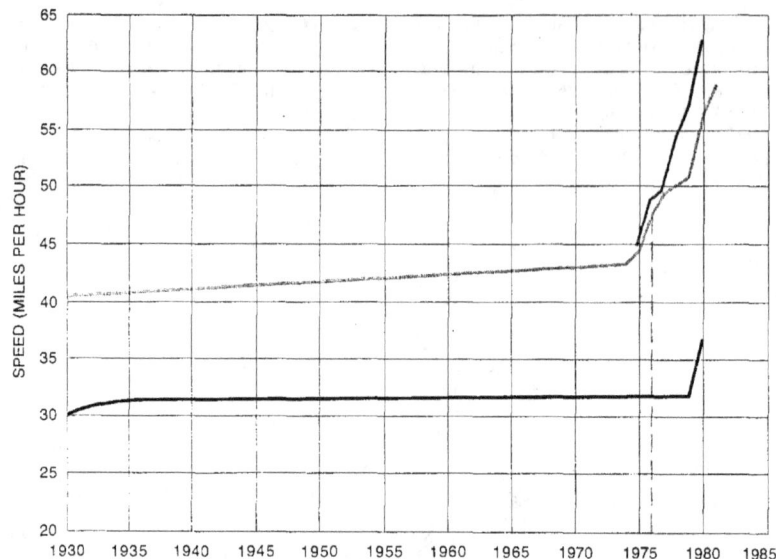

SPEED RECORDS of human-powered land vehicles have risen rapidly since the formation in 1976 of the International Human Powered Vehicle Association, which puts no restrictions on design. The time of the founding is indicated by the broken line. For many years before then the rules of the Union Cycliste Internationale, which banned streamlined vehicles from sanctioned bicycle competition, had kept speed records virtually unchanged. The curves represent records for multiple riders for 200 meters with a flying start (*color*), single riders under the same conditions (*gray*) and riders who pedaled for one hour at the maximum effort (*black*).

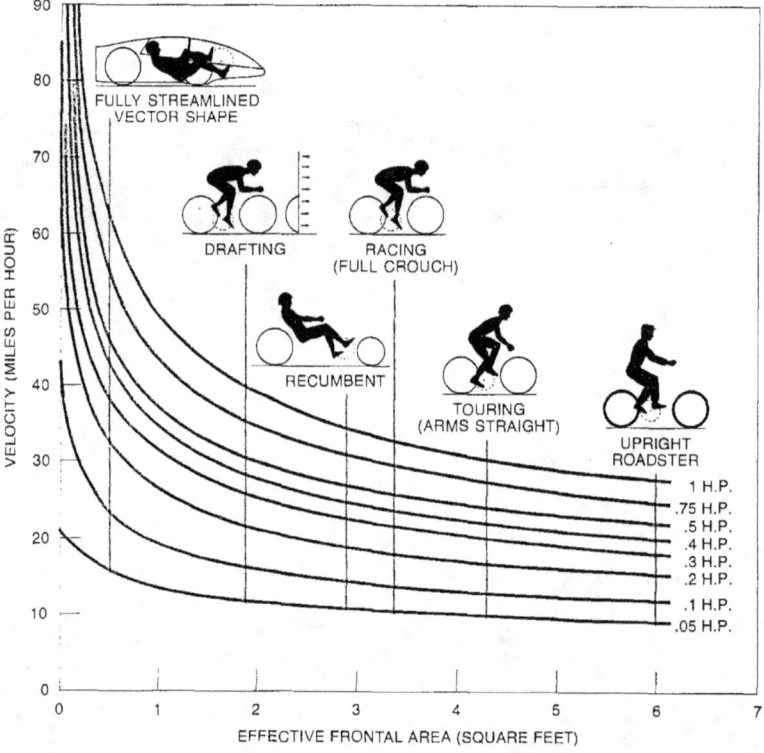

EFFECT OF STREAMLINING is to improve the performance of human-powered vehicles at all levels of power input. The upright roadster is the least streamlined vehicle, the Vector shape the most streamlined. Drafting means to follow closely behind another vehicle, here a bicycle. A good athlete can produce 1 h.p. for about 30 seconds, a healthy nonathlete for about 12 seconds. They can sustain an output of .4 and .1 h.p. respectively for about eight hours. The effective frontal area is the product of the drag coefficient and the projected frontal area.

MUSCLE POWERED BLIMPS

	DESCRIPTION		FORCES AT 20 M.P.H. (POUNDS)	AERODYNAMIC DATA			ROLLING-RESISTANCE COEFFICIENT	LEVEL-HORSE-POWER REQUIRED
				DRAG COEFFICIENT	FRONTAL AREA (SQUARE FEET)	EFFECTIVE FRONTAL AREA (SQUARE FEET)		
STANDARD BICYCLES	BMX (YOUTH OFF-ROAD RACER)	30-LB. BIKE, 120-LB. RIDER, KNOBBY TIRES, 20-IN. DIA., 40 P.S.I.	5.52 / 2.10	1.1	4.9	5.4	.014	146
	EUROPEAN UPRIGHT COMMUTER	40-LB. BIKE, 160-LB. RIDER, TIRES 27-IN. DIA., 40 P.S.I.	6.14 / 1.20	1.1	5.5	6	.006	140
	TOURING (ARMS STRAIGHT)	25-LB. BIKE, 160-LB. RIDER, CLINCHER TIRES, 27-IN. DIA., 90 P.S.I.	4.40 / .83	1	4.3	4.3	.0045	100
	RACING (FULLY CROUCHED)	20-LB. BIKE, 160-LB. RIDER, SEWUP TIRES, 27-IN., DIA., 105 P.S.I.	3.48 / .54	.88	3.9	3.4	.003	77
IMPROVED MODELS	AERODYNAMIC COMPONENTS (FULLY CROUCHED)	20-LB. BIKE, 160-LB. RIDER, SEWUP TIRES, 27-IN. DIA., 105 P.S.I.	3.27 / .54	.83	3.9	3.2	.003	73
	PARTIAL FAIRING (ZZIPPER, CROUCHED)	21-LB. BIKE, 160-LB. RIDER, SEWUP TIRES, 27-IN. DIA., 105 P.S.I.	2.97 / .54	.70	4.1	2.9	.003	67
	RECUMBENT (EASY RACER)	27-LB. BIKE, 160-LB. RIDER, CLINCHER TIRES, 20-IN. FRONT, 27-IN. REAR, 90 P.S.I.	2.97 / .94	.77	3.8	2.9	.005	75
	TANDEM	42-LB. BIKE, TWO 160-LB. RIDERS, CLINCHER TIRES, 27-IN. DIA., 90 P.S.I.	5.32 / 2.66 / 1.62 / .81	1	5.2	5.2	.0045	66
	DRAFTING (CLOSELY FOLLOWING ANOTHER BICYCLE)	20-LB. BIKE, 160-LB. RIDER, SEWUP TIRES, 27-IN. DIA., 105 P.S.I.	1.94 / .54	.50	3.9	1.9	.003	47
RECORD HOLDERS	BLUE BELL (TWO WHEELS, ONE RIDER)	40-LB. BIKE, 160-LB. RIDER, SEWUP TIRES, 20-IN. FRONT, 27-IN. REAR, 105 P.S.I.	.61 / .8	.12	5	.6	.004	27
	KYLE (TWO WHEELS, TWO RIDERS)	52-LB. BIKE, TWO 160-LB. RIDERS, SEWUP TIRES 105 P.S.I.	1.44 / .72 / 1.12 / .56	.2	7	1.4	.003	24
	VECTOR SINGLE (THREE WHEELS)	68-LB. BIKE, 160-LB. RIDER, SEWUP TIRES, 24-IN. FRONT, 27-IN. REAR	.51 / 1.02	.11	4.56	.5	.0045	29
	VECTOR TANDEM (THREE WHEELS)	75-LB. BIKE, TWO 160 LB. RIDERS, SEWUP TIRES, 24 IN. DIA.	.62 / .31 / 1.78 / .89	.13	4.7	.6	.0045	23
THEORETICAL LIMITS	PERFECT BIKE	NO ROLLING RESISTANCE, NO DRAG ON BIKE	3.07 / 0	.8	3.8	3	0	59
	DRAGLESS RIDER	ROLLING RESISTANCE INCLUDES RIDER'S WEIGHT	1.33 / .81	1.1	1.2	1.3	.0045	41
	PERFECT RECUMBENT	DRAG ON RIDER ONLY	.72 / 0	.6	1.2	.7	0	14
	PERFECT PRONE BIKE	DRAG ON SMALL BUT STRONG RIDER	.51 / 0	.6	.8	.5	0	10
	PERFECT PRONE STREAMLINER		.07 / 0	.05	1.4	.07	0	1
	MOTOR PACING	42-LB. BIKE, 160-LB. RIDER, MOTORCYCLE ROAD-RACING TIRES, 70 P.S.I.	0 / 1.21				VARIES WITH SPEED .006	23
	MOON BIKE	25-LB. BIKE, 160-LB. RIDER, 15-LB. SPACE SUIT	0 / .15				0 .0045	3

MUSCLE POWERED BLIMPS

GROUND, NO WIND		EFFECT OF HILLS	
ALL-DAY TOURING	MAXIMUM SPEED	STEADY SPEED UP	STEADY-SPEED COASTING
10.1	27.8	12.2	19.8
11.3	27.6	10.9	24
13.1	31.1	12.2	27.7
14.7	33.9	13	31.2
15	34.6	13	32.2
15.4	35.7	13.1	33.9
14.4	35.2	12.5	33.7
15.2	36.6	13	35.2
17.5	41	13.6	41.7
22.5	58.6	12.9	77.4
23.3	56.6	14	69.9
21.8	61.2	11.3	90.1
25.6	72.5	13	108.4
16.7	35.9	13.4	34.7
18.4	45.8	13.3	50.3
27.1	58.3	16.8	66.9
30.4	65.3	23.2	65.3
58.3	125.9	25.6	174.5
29.4	294	12.6	?
237.5	2,375	78.4	?

dents that do not involve an encounter with an automobile it is much safer, since the rider is closer to the ground (making falls less serious) and the feet are forward (making a head injury less likely in a fall). A problem is that a recumbent is hard to see on a road and so is perhaps more vulnerable to automobiles; the problem can be relieved somewhat by mounting on the vehicle a long, thin pole with a flag.

At the top of the expense ladder is a bicycle with a full fairing. The Vector Single, a one-rider version of the Vector Tandem, is the best example of a fully faired, enclosed, pedal-powered vehicle. (It is the machine portrayed on the cover of this issue of *Scientific American*.) According to Voigt, the vehicle is theoretically capable of reaching 61.7 m.p.h. with a 1-h.p. input, an increase of 28.2 m.p.h. over what has been done with a standard racing bicycle. A Vector Single costs about as much as a first-class racing bicycle.

In going up or down a hill a fully streamlined vehicle retains its advantage over a conventional bicycle. Although the Vector Single weighs about 80 pounds, compared with about 25 pounds for a standard bicycle, it can climb moderate hills as fast as or faster than the bicycle. With an input of .4 h.p. a bicycle can climb a 2.5 percent grade at about 16 m.p.h. and a 6 percent grade at about 11 m.p.h. With the same input the Vector can climb the two grades at 20.5 and 11 m.p.h. respectively.

Downhill the difference between the two machines is remarkable. The bicycle can descend a 2.5 percent grade at 29.5 m.p.h., the Vector at 54. On a 6 percent grade the bicycle can reach a speed of 39 m.p.h. and the Vector can exceed 100. Such potential speeds mean that if streamlined human-powered vehicles become common, careful attention must be given to the design of brakes and suspension and to the stability of the vehicle.

Since the aerodynamic drag force is proportional to the square of the relative velocity, head winds, tail winds and even crosswinds can drastically change both aerodynamic drag and the power requirements. For example, a bicyclist going at 18 m.p.h. in still air must increase his power output by 100 percent to maintain that speed against a head wind of 10 m.p.h. Usually a bicyclist confronting a head wind slows down and tries to maintain his customary leg force and pedaling cadence by shifting gears. This is one reason bicycles with multiple gears are desirable even for level country.

A tail wind makes the bicyclist go faster with his customary input of power. In general moving air will speed up or slow down a bicycle by about half the wind speed. When one bicyclist rides in the wake of another, the power requirements of the drafting rider are reduced by about 30 percent. The forward bicyclist creates an artificial tail wind.

The closer the rear bicycle follows the leader, the more pronounced the drafting effect is. One can think of the rear rider on a tandem bicycle as drafting extremely closely. Tandem riders use 20 percent less power per rider than two separate cyclists.

When the riders in a line of drafting bicyclists take turns in the lead position, the entire group can travel much faster than a single rider. In a pursuit race of 4,000 meters (almost 2.5 miles) a four-rider team can go about 4 m.p.h. faster than a single bicyclist. Typically a group of bicycle tourists of equal ability can travel from 1 m.p.h. to 3 m.p.h. faster than any rider alone. The larger the group is, the faster it should be able to travel (up to, say, a dozen riders).

Artificial winds created by passing automotive traffic can increase a bicyclist's speed from 1 m.p.h. to 3 m.p.h. for periods of about seven seconds. The larger the passing vehicle, the more substantial the effect. A steady stream of traffic can enable a bicyclist to sustain a speed from 3 m.p.h. to 6 m.p.h. higher than would otherwise be possible for a given energy input.

When a bicyclist rides directly in the wake of a motor vehicle, quite remarkable speeds can be attained. The practice is called motor pacing. On August 25, 1973, Allan V. Abbott, a physician in California, achieved a record of 138.674 m.p.h. motor pacing along a measured mile at the Bonneville Salt Flats in Utah. John Howard, a U.S. Olympic cyclist, is attempting to break Abbott's record and to achieve a motor-pacing speed in excess of 150 m.p.h.

Although the findings we have described are significant in their own right, one wonders if they will have any practical application beyond their effect on speed records. For a large fraction of the world's bicycle riders it seems unlikely that the work will have much immediate utility. For example, in the

PERFORMANCE OF HUMAN-POWERED VEHICLES is summarized. The numbers listed under forces for each vehicle represent air resistance and rolling resistance respectively. The five columns at the far right represent respectively the horsepower required at 20 m.p.h. as a percentage of the touring rider's performance; the all-day touring speed in miles per hour at an output of .1 h.p.; the maximum speed at an output of 1 h.p.; the steady speed in miles per hour up a 5 percent grade at an output of .4 h.p., and the coasting speed down the same grade.

many developing countries where the bicycle is the chief means of transportation most riders travel at about 7 m.p.h., often with a substantial load; aerodynamic drag becomes more important than other impediments to bicycle motion only at speeds above 10 m.p.h. Even here the work on aerodynamics makes a contribution. Without it designers would not know why they should largely ignore aerodynamics for slow-moving human-powered vehicles.

For bicycles intended for slow but sure progress it makes sense to decrease rolling resistance by improving tires and by paving roads. Designers should also reduce the bicycle's weight to facilitate climbing hills. The recent introduction of "mountain bikes" in the U.S. is a step in the direction of making lightweight bicycles durable enough for rugged or unpaved roads.

In several ways the knowledge gained by the recent research on the aerodynamics of human-powered vehicles can be directly useful. Although the standard bicycle is likely to be the predominant representative of the class for many more years because of its public acceptance, low cost, simplicity and mechanical reliability, it offers plenty of scope for innovation. For example, a light, simple and inexpensive front fairing will substantially improve the performance of the standard bicycle. The recumbent bicycle may come into greater use by commuters and tourists because of its efficiency and comfort.

A further application of the technology would be to fit a recumbent with a small, lightweight motor of low horsepower. The motor would serve mainly as an aid in accelerating from a stop and in climbing hills. Fitted also with as much streamlining as would be consistent with the need for ventilation and stability, the machine would be a true moped. (The machines now sold under that name are not really motor-pedal vehicles but merely underpowered motorcycles.)

The recent research has inspired inventors to develop several special-purpose human-powered vehicles. Paul Schöndorf, a professor of engineering at the Fachhochschule in Cologne, has built a series of easily pedaled, all-weather recumbent tricycles for the elderly and the handicapped. Similar vehicles would serve well in retirement communities. Douglas Schwandt of the Veterans Administration's Rehabilitation Engineering Research and Development Center at Palo Alto, Calif., has built hand-cranked tricycles and bicycles for paraplegics. William Warner, a paraplegic who once held the record for hand-powered vehicles in the races sponsored by the International Human Powered Vehicle Association, says a disabled person can propel such a vehicle much faster than a standard wheelchair and thereby can gain a new sense of freedom and mobility. (The present record of 25.09 m.p.h. was set in 1981 by Ascher Williams of the Palo Alto rehabilitation center.)

In principle a fully enclosed, streamlined human-powered vehicle could be quite useful in transportation. A rider could travel at speeds of from 20 to 30 m.p.h. in all kinds of weather. As such vehicles are now designed, however, they would not serve on the open road. They lack adequate ventilation, visibility, maneuverability and such safety features as lights and windshield wipers. Most of them are not easy to get into or out of.

To produce a practical vehicle of this kind would require an investment and an engineering effort comparable to that made in producing a new automobile. Even then the pedaled vehicle would not be safe in traffic that included a large number of motor vehicles. One must conclude that a fully enclosed human-powered vehicle will not be a practical form of transportation until fuel shortages remove most motorized vehicles from the roads or until special roadways are built for pedaled machines.

Far likelier is the development of lighter and more fuel-efficient automobiles employing much of the technology we have described. One of us (Malewicki) has already built such a vehicle, a single-passenger machine weighing 230 pounds. It holds records for fuel economy at the freeway speed of 55 m.p.h. with a gasoline engine (157.2 miles per gallon) and with a diesel engine (156.3 m.p.g.). The diesel record was set on a trip from Los Angeles to Las Vegas, during which the average speed was 56.3 m.p.h. A trend toward such vehicles could help to extend fuel resources and ironically might postpone the time when the human-powered vehicle will have come fully into its own.

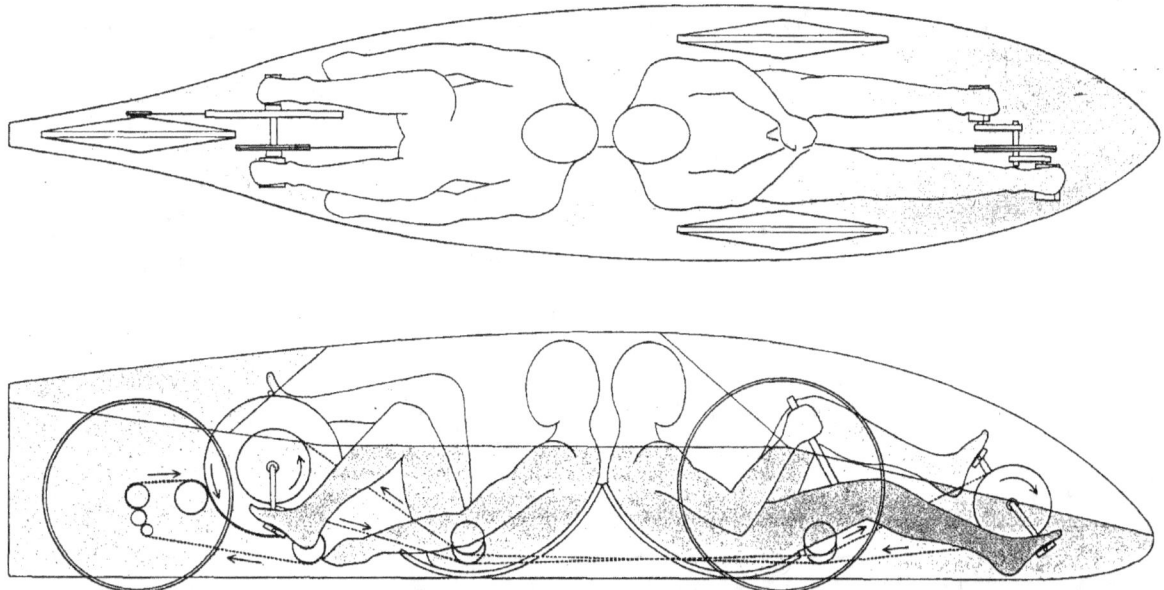

VECTOR TANDEM is shown in plan and elevation. It is a companion vehicle to the Vector Single portrayed on the cover of this issue of SCIENTIFIC AMERICAN. The Tandem, receiving an input of a bit more than 1 h.p. from each of its two riders, who are positioned back to back, set the speed record of 62.92 m.p.h. for 200 meters in 1980. (The riders had a flying start of more than one mile.) Traveling on an interstate highway in California later that year the Vector Tandem managed an average speed of 50.5 m.p.h. on a trip of 40 miles.

MUSCLE POWERED BLIMPS

SCIENTIFIC AMERICAN

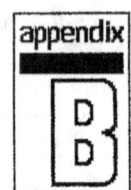
appendix B

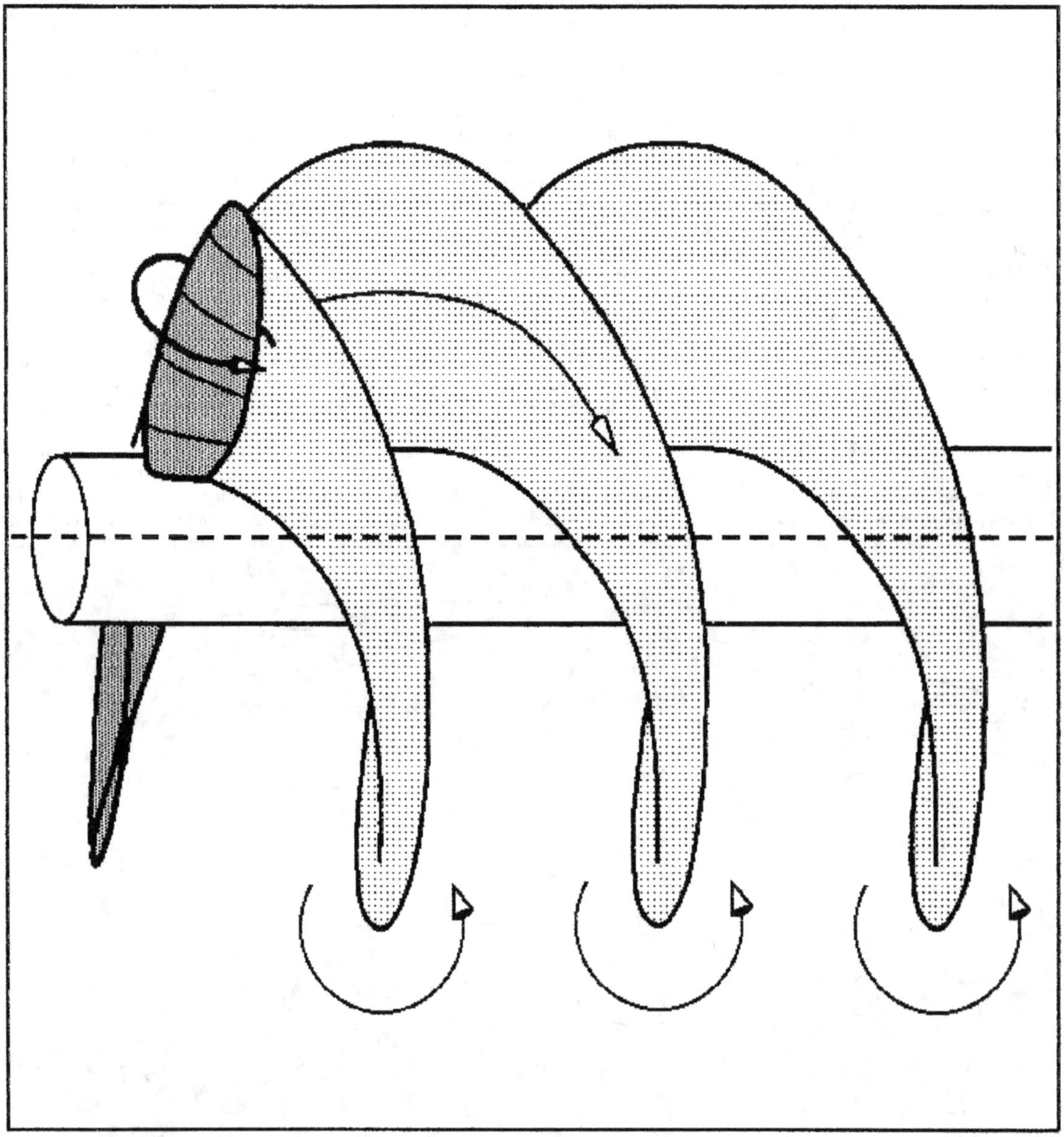

THE SCREW PROPELLER

$2.50

July 1980

The Screw Propeller

Ship screws and aircraft propellers look quite different, but both are designed according to the circulation theory of lift. The theory is the traditional one developed to account for the lift of an airfoil

by E. Eugene Larrabee

In 1845, early in the history of steam navigation, the effectiveness of the screw propeller was established when the steam sloop *Rattler,* equipped with such a propeller, towed the steam sloop *Alecto,* equipped with paddle wheels, stern first at a speed of nearly three knots in a tug-of-war staged by the British Admiralty. The closely matched ships were driven by slow-turning steam engines of about 200 horsepower. The test was significant in that it demonstrated the validity of screw-propeller propulsion in the face of the intuitively more understandable concept of the paddle wheel.

The screw propeller has remained the most generally useful means for driving watercraft of all types and sizes. The chief exceptions are a few special-purpose craft such as shallow-draft or hydrofoil boats. And when it comes to moving a piston-engine airplane through the air, there is no alternative to the propeller. The propeller has also been retained on turboprop aircraft powered by gas turbines. Most commercial airliners, however, are now driven by turbofan engines, in which the propeller goes by another name. In a turbofan engine the high-velocity gas output from the fuel-combustion chamber flows through a power turbine that drives a "fan," or axial compressor, at the front of the engine. The fan is really a many-bladed screw propeller designed to operate within the engine case. The fan compressor provides the air for combustion and also creates a low-velocity, high-volume flow of air that bypasses the combustion chamber and mixes with the hot exhaust of the power turbine to propel the airplane. In the turboprop engine the power turbine turns an external visible airscrew that provides most of the thrust. As we shall see, the turboprop, by virtue of its efficiency, may be about to rise again.

Anyone who has looked closely at a motorboat propeller and a typical airplane propeller is aware that the blades of a marine screw fill a much larger fraction of the total disk area, the area swept out by the propeller blades, than the blades of an airscrew do. The aeronautical engineer would say that the marine screw has greater "solidity." An extreme contrast can be seen if one compares the marine screw, or "wheel," of an ultra-large crude carrier (oil tanker) with the airscrew of the *Gossamer Albatross,* the ultralight aircraft that Bryan Allen pedaled across the English Channel on June 12 of last year. The propellers of the largest tankers have five or six blades, a diameter of up to nine meters, turn at about 95 revolutions per minute and can absorb 45,000 h.p. to drive a ship of more than 500,000 deadweight metric tons at 16 knots (8.23 meters per second). The twin screws of the 31-knot *Queen Elizabeth 2,* each of which absorbs 55,000 h.p., are appreciably smaller, turn much faster and have an even greater solidity than the screws of the largest tankers.

The *Gossamer Albatross* propeller has two slim blades and a diameter 70 percent as large as the screws of the *Queen Elizabeth 2:* when the 4.1-meter propeller is turned at 95 r.p.m., it absorbs .25 h.p. to drive an airplane with a flying weight of 96 kilograms at 5.4 meters per second. Marine screws and airscrews are both the result of similar computer analyses that account for the fluid density, the interaction of the flow fields, the hydrodynamic or aerodynamic load on the blades and finally the practical limitations imposed by the vehicle and its mode of operation. Why are the geometries of the two kinds of propeller so different?

It is not because air is a gas and water is a liquid. Strangely enough, the *Gossamer Albatross* propeller comes much closer than the ship screw to the ideal form of a propeller of highest efficiency adapted to operation in a nearly incompressible fluid of low viscosity, that is, a fluid such as water. Its shape conforms to theoretical considerations set forth by Albert Betz and Ludwig Prandtl of the Kaiser Wilhelm Institute for Fluid Dynamics Research at Göttingen in 1919 and refined by a visiting British scholar, Sydney Goldstein, in 1929. Their ideas came out of the revolution in theoretical hydrodynamics at the beginning of the century, when Prandtl invented boundary-layer theory, which explained the drag, or resistance, of streamlined bodies, and when W. M. Kutta of Germany and Nikolai Joukowsky of Russia independently invented the circulation theory of lift, which explained the lift of wings and propeller blades.

Until then theoretical hydrodynamics, which had been invented to study the resistance of ship hulls in motion, had been largely an academic discipline concerned with the mathematical flow patterns that could be formed by various combinations of flow fields. Although the streamlines of some of the combined mathematical flow fields could be made to resemble the flow around a ship hull, Jean Le Rond d'Alembert had shown in 1742 that the theoretical resistance of such a flow on the mathematical hull form was exactly zero. Moreover, if the hull form was mathematically yawed (turned at an angle to the flow field), the hydrodynamic force perpendicular to the direction of hull motion was also zero. Since real hulls do not exhibit zero response to real flows, these findings became known as d'Alembert's paradox.

In the absence of a useful theory 19th-century naval architecture became an experimental science and marine propellers were developed intuitively. The propeller was likened to a machine screw that advances as it turns in a threaded hole. Unlike metal, however, water yields under the thrust of the propeller. Hence the "effective pitch," or the distance traveled in one revolution, is less than the "geometric pitch" of the propeller by an amount called the slip. In 1865 the Scottish engineer William J. M. Rankine developed an "actuator disk" theory of the propeller, later improved by William Froude, in which the

MUSCLE POWERED BLIMPS

PROPELLER OF "QUEEN ELIZABETH-2" is one of two screws that together absorb 110,000 horsepower in driving the liner of 67,000 gross tons At a top speed of 31 knots. The propeller is 5.8 meters in diameter and generates maximum thrust when it is turning at 174 revolutions per minute. The screw's diameter is sharply constrained by the requirement that it not project below the keel or beyond the vessel's beam. In order to achieve the required thrust within the limited diameter the total blade ares is 89.5 percent as large as the area of a solid disk of the same diameter. In marine terminology the ratio of blade area to disk area is called the expanded area ratio; in aeronautics the same ratio is called solidity The *Queen Elizabeth-2* propeller was designed and made by SMM Propeller Ltd. of Liverpool.

PROPELLER OF "GOSSAMER ALBATROSS," The ultralight aircraft Bryan Allen pedaled across the English Channel on June 19, 1979, bears only a remote resemblance to the screw of the *Queen Elizabeth-2* shown at the top of the page. Both propellers, however, act us "pushers and are among the most efficient of their type ever designed. The airscrew of the *Gossamer Albotross* is 4.1 meters in diameter, rotates at 95 p.m. and absorbs .8 hp. at takeoff and .25 h.p. in cruising flight. The propeller was computer-designed with the author's algorithms to have minimum energy in its vortex wake and minimum friction loss in slightly climbing flight. The

Scientific American, July 1980 135

slip could be explained as the consequence of a simple increase in the momentum of the fluid in the propeller slipstream, an increase proportional to the thrust of the propeller.

In this theory the slip is half the ultimate increase in slipstream velocity (which depends on the thrust per unit disk area and the speed of the ship) multiplied by the time for one propeller revolution. The geometric pitch must be increased accordingly when one is trying to arrive at the effective (or desired) pitch, which is the ratio of the ship speed in feet per minute to the shaft speed in revolutions per minute. Although all of this is almost true experimentally, the actuator-disk theory gave no insight into why the blades should develop hydrodynamic thrust loads and torque loads in the first place. It did, however, give a precedent for the idea of "inflow" velocity, which would later be identified with the "induced" velocity of wing theory, and it associated an excessive slip (and a high disk loading) with reduced efficiency.

An explanation of thrust and torque loads awaited the development of the "vortex," or circulation, theory of lift by Kutta and Joukowsky in the period between 1902 and 1911. In Joukowsky's version of the theory the ideal flow of an inviscid fluid around a circular cylinder is conformally "mapped," or transformed, into the corresponding flow around an airfoil with a sharp trailing edge. Vorticity, or circulation, is added to the flow around the cylinder by introducing an idealized "bound," or "line," vortex of arbitrary strength coincident with the center of the cylinder. The strength of the vortex is adjusted mathematically until the flow around the cylinder stagnates at a singular point that coincides in the transformation with the trailing edge of the airfoil.

The hypothetical circulation of inviscid fluid around a cylinder of infinite length superficially resembles the flow of real fluid around a rotating sphere with a roughened surface such as a tennis ball or golf ball moving with backspin. Lift is produced as a consequence of the acceleration of the air flowing over the top of the ball and the deceleration of the air flowing under the ball. In accordance with a well-known theorem of the 18th-century Swiss mathematician Daniel Bernoulli, the fluid in the accelerated flow has a lower static pressure than the fluid in the decelerated flow, so that the ball is pushed upward from below. (A tennis ball hit with topspin receives a corresponding downward push from above.) Although ideal inviscid flow fields satisfy Bernoulli's theorem, there can be no lift without circulation. As a result the usual secondary-school explanation of airfoil lift in terms of Bernoulli's theorem is entirely inadequate.

In that explanation lift is created because the air passing over the curved upper surface of a typical airfoil must travel farther, and therefore faster, than the air passing under the flatter lower surface. Hence the pressure above the wing must be less than the pressure below the wing, resulting in lift. Actually lift can be generated by a perfectly symmetrical airfoil that is tilted upward at a suitable angle of attack. How does the airstream know how to divide at the

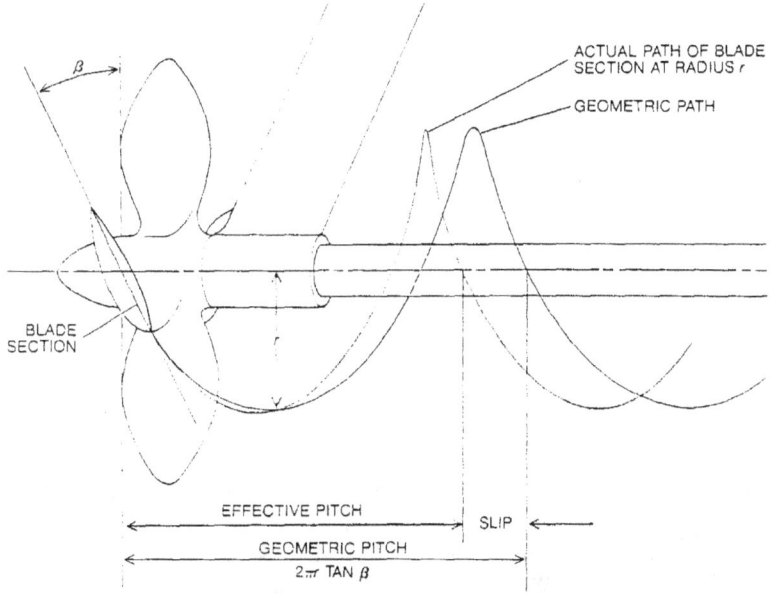

SLIP OF MARINE PROPELLER was recognized early in the history of marine screws. The distance actually traveled by a ship during one revolution of the propeller, the "effective pitch," is less than the geometric pitch determined by the propeller's blade angle. William J. M. Rankine and William Froude explained that slip corresponds to the increase in axial momentum imparted to slipstream by propeller. They showed that excessive slip leads to low efficiency.

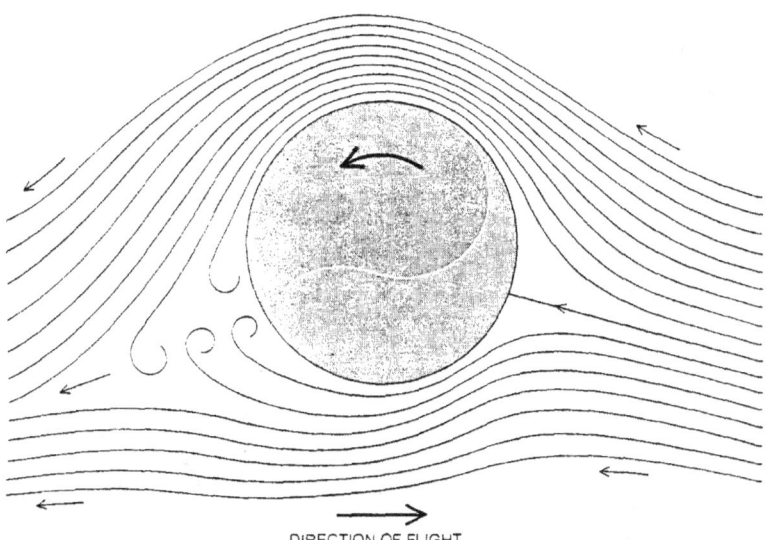

TENNIS BALL HIT WITH BACKSPIN mimics an airfoil by distorting the flow field in a way that creates aerodynamic lift. Perhaps the earliest explanation of the flight of a "cut" tennis ball is attributed to Lord Rayleigh. Because of backspin the air flowing over the top of the ball is accelerated to the rear and the air flowing under the ball is retarded. The effect is enhanced by the ball's fuzzy surface. According to Bernoulli's theorem, a statement of energy conservation, the pressure in the accelerated fluid above the ball, compared with the pressure in the retarded fluid below the ball, must drop. The resulting imbalance of forces pushes the ball upward.

MUSCLE POWERED BLIMPS

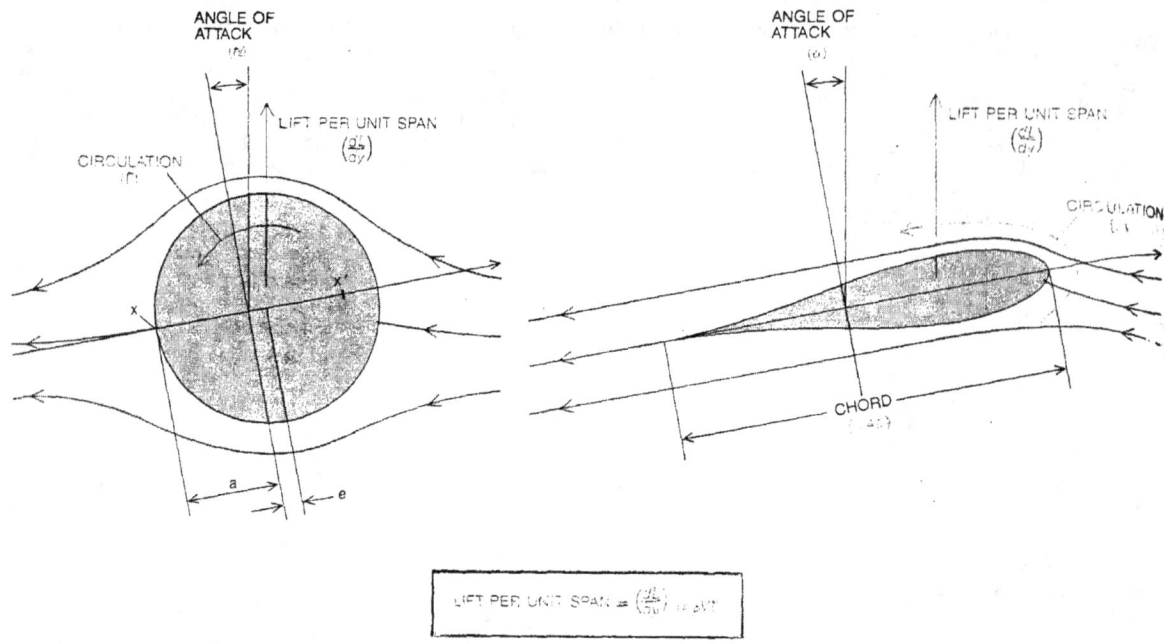

VORTEX THEORY OF LIFT was invented independently between 1902 and 1911 by W. M. Kutta of Germany and Nikolai Joukowsky of Russia. In Joukowsky's formulation, depicted here, an inviscid flow around a cylinder of infinite length is "mapped" mathematically into the flow around an airfoil. The flow can be symmetrical, as is shown. Joukowsky added to the inviscid flow an idealized vortex flow coincident with the center of the cylinder. The strength of this "bound" vortex, or circulation, Γ, is chosen so that it gives rise to a stagnation in the flow at the aft singular point x on the cylinder. When this circulation is mapped onto the airfoil, it aligns the flow field with the sharp trailing edge. Under these conditions the circulation creates a lift on both the cylinder and the airfoil. The lift per unit of span is equal to $\rho V \Gamma$, where ρ is the fluid density (in this case the fluid is of course air) and V is the velocity of the remote, or undisturbed, stream. The thickness of the airfoil is controlled by the value that has been assigned to the offset dimension e. The airfoil angle of attack, α, corresponds to the misalignment of the two singular points of transformation, designated x and x', with respect to the remote stream direction.

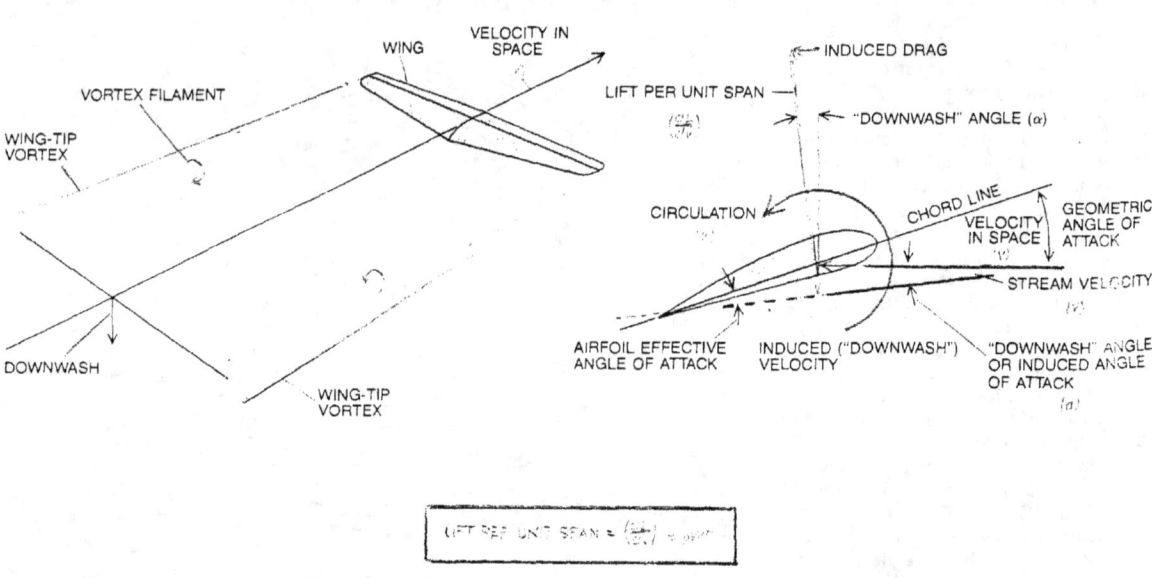

CONCEPT OF TRAILING VORTEX SHEET was introduced by Ludwig Prandtl to account for the flow around wings where the lift and the bound vorticity must fall to zero at the wing tips. The vortex sheet (*left*) is made up of filaments, or infinitesimal line vortexes, aligned approximately with the direction of flight and increasing in strength toward the wing tips, where they roll up to form tip vortexes. The collective velocity field of the vortex filaments imparts a "downwash," or induced velocity, to the vortex sheet as part of the roll-up process. As is shown at the right, the downwash velocity (w) is added vectorially to the flight velocity in space (V) to produce a local stream velocity (W) that is misaligned with the flight velocity by the induced angle of attack (α). In order to produce a specified lift the geometric angle of attack of the wing section must be larger than the airfoil's effective angle of attack by an amount equal to the induced angle. As a result the lift vector (dL/dy) will be rotated aft by an amount equal to the induced angle, thereby giving rise to an induced drag.

leading edge and reunite in the trailing edge so that the flow over the upper surface of the wing will on the average have a higher velocity and a lower pressure than the flow under it.

The angle of attack is therefore curtail, but there is no satisfactory physical theory to explain why the flow divides precisely the way it does. Joukowsky's adjustment of the vortex strength to produce stagnation at the aft singular point of the transformed cylinder that is the airfoil has the effect of aligning the flow held with the trailing edge of the airfoil, a plausible assumption called the Joukowsky hypothesis or the Kutta Joukowsky hypothesis. The effect of aligning the flow at the trailing edge is to misalign, or offset the flow where it divides at the leading edge of the airfoil in such a way that circulation and lift result. The leading edge should be well rounded, as Joukowsky's mathematical airfoils are.

The angle of attack of the airfoil can be varied by tilting the axis connecting the fore and aft singular points of the transformation with respect to the direction of the remote flow field, that is, its direction before it has been influenced by the airfoil. The airfoil camber, or centerline curvature, can be varied by offsetting the radius joining the center of the cylinder to the aft singular point from the axis of the singular points in the transformation. Each of these misalignments involves a change in the strength of the circulation, or bound vortex, and gives rise to corresponding changes in the lift of the airfoil. Wind tunnel tests 0f Joukowsky airfoils done by Joukowsky and his associates in Russia and by Prandtl and his associates in Germany just before World War 1 verified the predictions of the theory of small angles of attack and small amounts of camber. All modern airfoil theories incorporate Joukowsky's results as a special case.

Although Joukowsky's theory explains the lift of wing sections in two dimensional flow (that is, in the absence of spanwise components in the flow velocity), it presents difficulties when it is applied to the flow around wings of finite span where spanwise flow exists and where lift and the bound vorticity must all go smoothly to zero at the wing tips. This problem was solved by Prandtl, Betz, Max Munk and C. Wieselberger at Gottingen shortly before or during World War 1. They introduced the concept of a trailing vortex sheet made up of infinitesimal line vortexes roughly aligned with the direction of flight. The strength of each of the trailing vortexes is proportional to the local spanwise gradient of the bound vorticity associated with the local value of lift of the airfoil section on Joukowsky's model.

The trailing vortexes are particularly

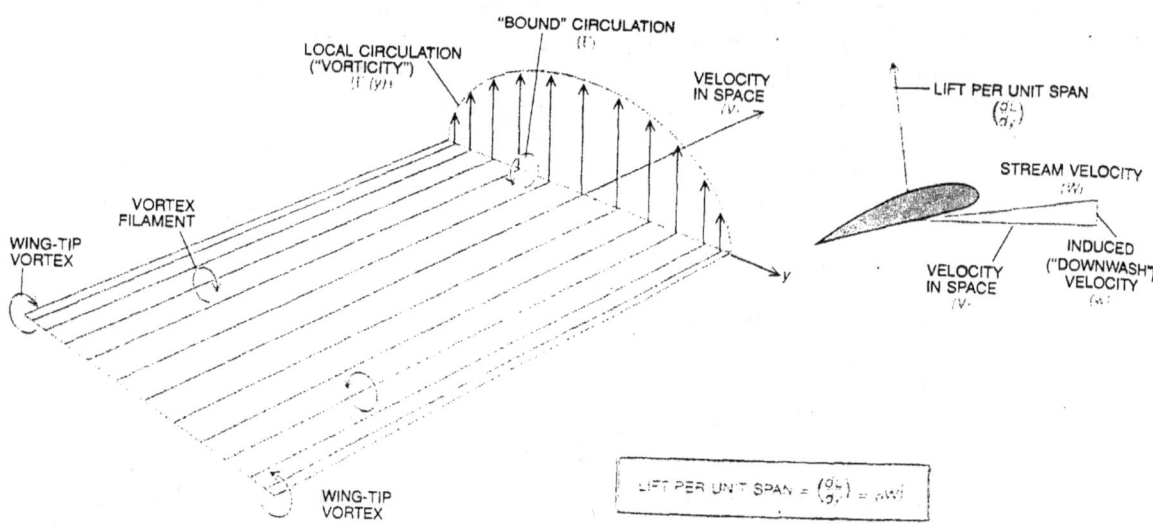

OPTIMUM DISTRIBUTION OF LIFT, corresponding to the minimum induced drag, is achieved when the lift and the "bound" circulation vary elliptically from wing tip to wing tip. With that distribution the velocity field induced by the trailing-vortex system will convect the vortex sheet downward with a uniform downwash velocity across the entire span. In this condition a minimum amount of kinetic energy is transferred to the vortex wake for a given amount of lift. The minimum-loss condition was found by Prandtl's colleague Max Munk.

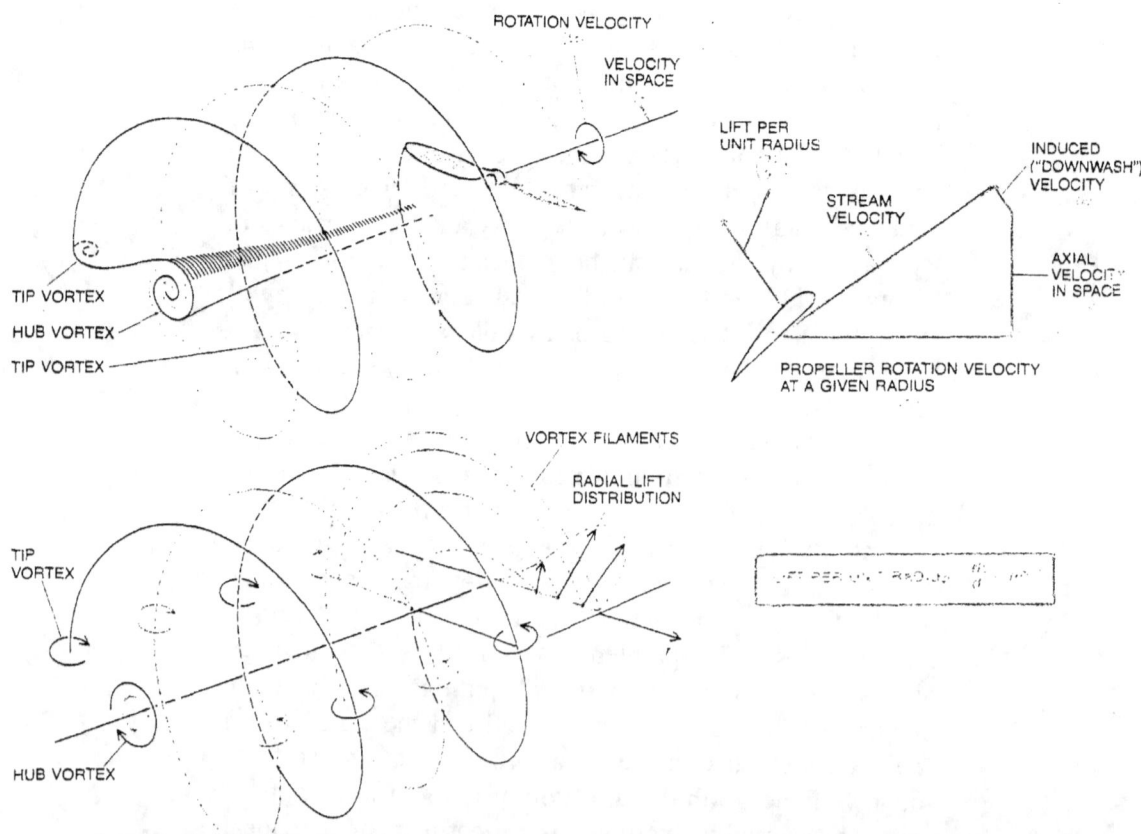

VORTEX THEORY OF PROPELLERS is like the vortex theory of wings except that the trailing vortex sheets of propellers are helicoidal surfaces (*top*). The sheets roll up into as many tip vortexes as there are blades in the propeller and into one central hub vortex equal in strength to the sum of the tip vortexes. The vectors of induced velocity (w) are perpendicular to the surface of the helicoidal sheets and correspond in magnitude to the slip in the Rankine-Froude momentum theory of marine screws. For minimum induced loss each helical vortex filament (*bottom*) should move perpendicularly to itself in such a way that the entire vortex-sheet array appears to move axially with a constant "displacement" velocity. The vortex-sheet motion for the minimum loss was discovered by Prandtl's colleague Albert Betz.

strong near the edges of the sheet, at the wing tips, where the gradient tends to reach infinity. The trailing vortex sheet rolls up around its intense outer edges, forming two distinct "tip vortexes" (which, however, were not accounted for in the primitive theory). The roll-up process is substantially complete several wingspans behind the trailing edge, where the tip vortexes are spaced less than a wingspan apart. When the humidity of the upper atmosphere falls within a certain range, the water vapor from the exhaust of an aircraft's engines will condense within the cores of the tip vortexes and make the vortexes visible as condensation trails.

The collective velocity field of the vortexes within the vortex sheet gives rise to a "downwash," or induced velocity, that tends to convect the sheet downward at every stage in the roll-up process. In particular, if the lift of the wing is distributed elliptically in the spanwise direction, the vortex sheet (before it rolls up) is convected downward with a uniform velocity across its entire width. The kinetic energy associated with the trailing vortex flow field is unrecoverable, but the losses are minimized for a given lift, flight speed and wingspan if the lift is elliptically distributed. This important result is attributed to Munk.

Joukowsky's model for airfoil lift in two-dimensional flow can be incorporated into a vortex theory for wings of finite span by calculating the magnitude of the downwash induced by the trailing-vortex system. The vector sum of the downwash velocity and the flight velocity creates an induced angle of attack. The necessary geometric angle of attack with respect to the undisturbed air is obtained by adding the induced angle of attack to the theoretical angle of attack required for a wing of infinite span in two-dimensional (Joukowsky) flow.

The vectors of wing lift also are rotated to the rear through an angle equal to the induced angle of attack, thereby creating an induced drag. Finally, the rate of loss of energy corresponding to the product of the induced drag and the flight velocity exactly balances the kinetic energy continuously added to the vortex system. There is in addition a frictional drag on the wing sections due to the operation of viscosity in the boundary layers adjacent to the wing surfaces, an effect that was also described by Prandtl. The predictions of the vortex theory of wings were confirmed by wind-tunnel experiments at Göttingen during World War I.

From here it is only a short step to a vortex theory of propellers. The essential difference is that the trailing vortex sheets of a propeller are initially helicoidal. Eventually they roll up into as many tip vortexes as there are blades and into one central hub vortex of opposite rotation whose strength is equal to the sum of the strengths of the individual tip vortexes. The collective velocity field of the helicoidal vortex sheets is identical with the "tailrace," or slipstream, of the running propeller. And just as there is a single vortex-sheet motion that minimizes the induced losses of the lifting wing, so there is a single vortex-sheet motion that minimizes the induced losses of a thrusting propeller.

This motion was discovered by Betz, and the corresponding (but approximate) radial distribution of bound vorticity for a Joukowsky model of optimum distribution of blade lift was calculated by Prandtl. They published their results in 1919. Goldstein joined the Göttingen group in the 1920's and defined the errors in Prandtl's approximate circulation distribution for his doctoral dissertation in 1929.

It is perhaps fair to say that even aero-

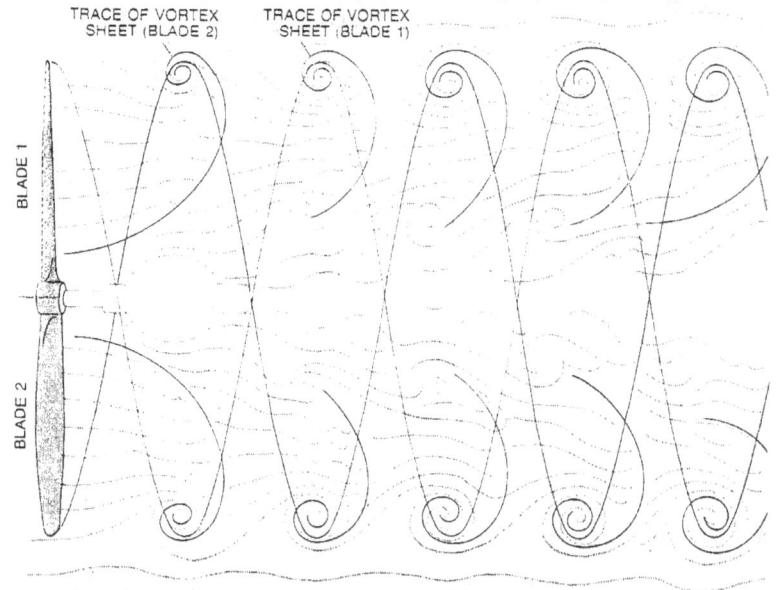

VORTEX WAKE OF TWO-BLADED AIRSCREW, a model-airplane propeller (*top*), was photographed in a special wind tunnel in the late 1950's by F. M. N. Brown of the University of Notre Dame. The photograph shows how smoke filaments in a vertical array are twisted up into a series of tip vortexes uniformly spaced at the pitch of the propeller. The vortex at the hub of the propeller is disrupted by the drive-shaft housing. Diagram at the bottom shows how curved discontinuities correspond to sections through helicoidal vortex sheets shed by blades.

The Screw Propeller

nautical engineers have had difficulty assimilating the Betz-Prandtl-Goldstein contributions precisely because the vortex sheets of a propeller, and their geometric development in space, are hard to visualize in three dimensions. The propeller is, after all, a peculiar kind of wing. Unlike a wing, which in straightaway flight meets the air everywhere at the same average velocity and so has a common angle of attack from root to tip, the propeller meets the air at a different velocity at each radius. That is why a propeller blade is designed with a twist. Near the hub, where the axial, or forward, velocity of the blade section is high compared with the circumferential velocity, the chord line, or tilt, of the blade departs only slightly from the direction of flight. At the blade tip, where the circumferential velocity is high compared with the forward velocity, the chord line may be twisted almost at right angles to the line of flight.

As we have seen, the velocity field of the trailing vortex sheet creates a downwash that is perpendicular to the direction of the airfoil's motion through undisturbed space. Near the hub of the propeller, therefore, the downwash bends backward only slightly from the line of flight. In other words, the vector of induced velocity near the hub has a high rotational component and only a small axial component. Accordingly the blade sections near the hub contribute only a small part to the propeller's total forward thrust. Near the tip of the blade, in contrast, the induced velocity has a high axial component and a much smaller rotational component. As a result the thrust per unit radius for a blade of constant section would increase from hub to tip except near the tip, where it must fall to zero "elliptically."

It might be thought that since the outer part of the blade does more useful work than the inner part, the outer part should be emphasized in the design of the propeller. One can imagine, for example, a propeller whose blades expand in width from hub to tip. Such a geometry (even if it were structurally attainable) would be relatively inefficient. The propeller counterpart of the spanwise uniform downwash velocity for an efficient wing is achieved when the induced velocity at any radius of the propeller blade is equal to half the vortex-sheet velocity at the same radius, and when the vortex-sheet velocities, axial and rotational, vary with the radius in such a way as to present the appearance of a uniform axial "displacement" velocity. A rotating barber pole with helical stripes, for example, has a finite axial displacement velocity but no real physical axial velocity.

In general the propeller geometry of highest efficiency, known as the geometry of minimum induced loss, is that of a propeller whose blades have a maximum chord near 30 or 40 percent of their radius and taper to a pointed tip. Such blades were a common feature of the aluminum-alloy propellers made in the late 1920's and early 1930's. As engines became more powerful and were supercharged to reach higher altitudes it became commonplace to increase the blade area toward the tip in order to absorb the higher power without increasing the propeller radius. Larger diameters were ruled out because tip speeds were already transonic, that is, moving at roughly the speed of sound, a regime for which no satisfactory blade designs were known. In extreme cases the tips were even made square. Such departures from the geometry of minimum induced loss led to small losses in propeller efficiency that were not noticed by the large majority of aeronautical engineers of the day, most of whom probably had not read, or perhaps had not understood, Goldstein's work (or the commentaries on it by Hermann Glauert and later by Theodore Theodorsen). Only when the designer needs every iota of efficiency is minimum-induced-loss design essential.

My own contribution to propeller theory has been to develop a "radially graded" momentum theory that lends itself to computer implementation and that is consistent with the Betz-Prandtl form of propeller vortex theory. All propeller theories seek to calculate the induced velocity of the propeller-blade elements that is due to the trailing-vortex system. In the case of propellers with minimum induced loss the half-displacement velocity corresponds to the slip velocity of actuator-disk theory of Rankine and Froude. As in wing theory, the axial and rotational components of the induced velocity (called the inflow velocity in propeller theory) require an increase in blade angle for a specified thrust together with a backward rotation of the blade-lift vectors, thereby increasing the torque needed to turn the propeller. The required increment corresponds to an induced torque. The vortex theory of propellers is hence entirely consistent with the early momentum theories of marine propellers and accounts for the details of blade loading that correspond to a specific propeller geometry.

Propellers, like wings, also have "profile," or friction, losses owing to the action of viscosity in the air flowing near the airfoil surface: the boundary layer. These losses are minimized by operating the blade elements at angles of attack where the ratios of drag to lift are low and, if possible, by choosing helix angles equal to 45 degrees minus half the angle whose tangent is the lift-to-drag ratio of the blade in two dimensions. Reducing the friction losses therefore calls for fairly high "advance ratios" (ratios of forward speed to propeller-tip speed of about .7) and a small number of blades, so that the solidity (or blade-to-disk area) can be concentrated in a few wide blades that will operate at high Reynolds numbers and correspondingly low skin friction. (The Reynolds number is a ratio of the momentum forces to the viscous forces in a fluid and helps to determine the character of the boundary layer and its resistance to separation.) Induced losses, on the other hand, are minimized by increasing the number of blades and lowering the advance ratio in order to decrease the spacing between individual vortex sheets and to make them more nearly perpendicular to the propeller axis. The objective is to make the slipstream velocity field as uniform as possible. The requirements for low induced losses and low friction losses in a propeller are therefore opposed to each other; propellers of the highest efficiency require an optimum balance of the two.

By the use of the approximate analytic Betz-Prandtl circulation distribution I have been able to write down simple algorithms, or computational procedures, that enable a propeller designer to determine the geometry of a propeller of minimum induced loss after accepting whatever design constraints may set a floor on reaching the lowest possible induced loss. (For example, if the design necessitates a two-blade propeller,

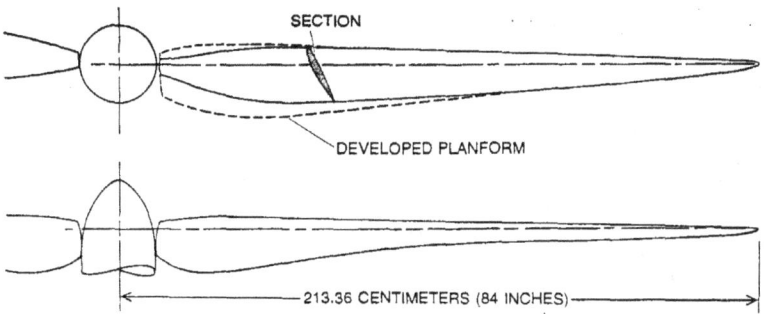

"CHRYSALIS" PROPELLER is shown from the front (*top*) and the side (*bottom*). The "developed planform," indicated by the broken outline, is the geometry the blade would have if it were untwisted. The maximum chord is at 30 percent radius. The propeller diameter of 4.27 meters is about 4 percent larger than that of *Gossamer Albatross*. Both propellers were designed according to the author's algorithms for minimum induced loss. *Chrysalis* was designed for an input of .5 h.p. (373 watts), a propeller-rotation rate of 135 r.p.m., a flight velocity of 4.88 meters per second (10.9 miles per hour) and therefore an "advance ratio" (forward speed divided by blade-tip speed) of .1617. Because of their high tip speed (6.26 times flight speed) propellers of this type would be unsuitable for aircraft flying more than 50 meters per second (112 m.p.h.).

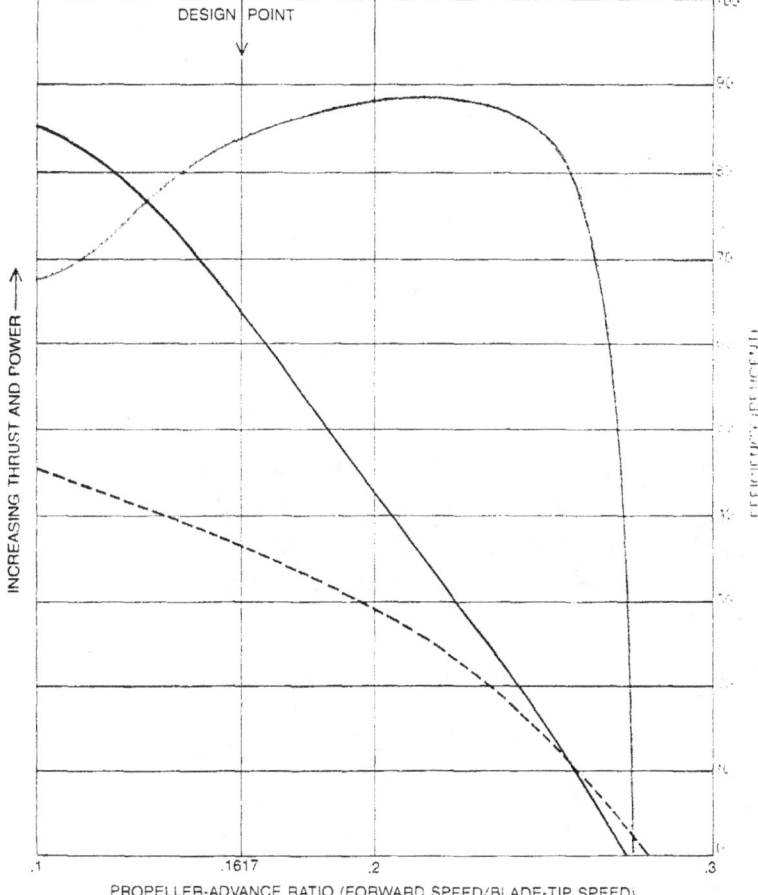

CALCULATED PERFORMANCE of the *Chrysalis* propeller shows how the coefficients of thrust (*solid black curve*), power (*broken black curve*) and overall efficiency (*color*) vary with the propeller advance ratio. The minimum induced loss for the design disk loading occurs at the design point where the propeller efficiency approaches 83 percent. The efficiency continues to improve at higher advance ratios (equivalent to turning the propeller more slowly at a fixed flight speed) even though the induced losses are no longer minimized; they simply become less as the load on the propeller is reduced and the average axial velocity of the slipstream is decreased. Coefficients of thrust and power are higher when the propeller is turning faster than at the design point (*curves to the left of the design point*) and fall off rapidly with slower turning.

for whatever reason, the designer cannot explore the possible advantages of adding more blades.) These algorithms thus yield a propeller of the highest possible efficiency that will develop a given thrust or absorb a given amount of power for a specified diameter, number of blades, flight speed, shaft speed and air density. The algorithms also account for a wide range of lift-to-drag ratios for blade elements of different contours in order to achieve desired lift coefficients at specified design points. Since the calculations give the propeller efficiency not only at the design point but also under flight conditions away from the design point, they can be used by the designer to change his design point.

It has been my good fortune to take part in a practical demonstration of the validity of what may properly be called a classical theory of propeller aerodynamics, first published in 1919 (a year before I was born) and practically forgotten by 1942, when I began my professional career at the Buffalo, N.Y., plant of the Curtiss-Wright Corporation. I owe this demonstration to enthusiastic students at the Massachusetts Institute of Technology who wanted to build a pedal-driven airplane in January, 1979, and who called on me for propeller-design concepts I had organized mainly for presentation at a symposium.

The students were fired by the dream (remote, to be sure) of winning the prize of £100,000 offered by Henry Kremer, a British industrialist, for a human-powered airplane capable of crossing the English Channel. (An earlier Kremer prize of £50,000 for the first human-powered airplane capable of completing a figure-eight course around two pylons half a mile apart had been won in August, 1977, by *Gossamer Condor*, an ultralight craft designed by Paul MacCready, president of AeroVironment Incorporated of Pasadena, Calif.)

As a first step my students Hyong Bang, Robert Parks and Harold Youngren coded up a version of my propeller algorithms for machine calculation and applied them to the redesign of the propeller for an eighth-scale, free-flying radio-controlled model of a biplane they planned to name *Chrysalis*. The model, with a wingspan of 2.74 meters, was powered by the motor from a Polaroid SX-70 camera that turned a propeller with an 11-to-one reduction gear. The model was constructed in one week in February. Compared with an intuitively designed propeller, with which the model would barely maintain level flight with fully charged batteries, the theoretically correct propeller, matched to the airframe and power plant, enabled the model to fly three and a half figure eights around pylons 20 meters apart in the M.I.T. gymnasium before the batteries ran down.

The second success for the revived

Betz-Prandtl theory came when the same students designed a new propeller for the *Gossamer Albatross* monoplane, which MacCready's group had built for its own attempt at the cross-Channel prize. Bryan Allen, a young professional bicyclist who weighed 137 pounds, had been able to keep *Gossamer Albatross* aloft only 17 minutes when it was equipped with its original propeller. With the new M.I.T. propeller, in combination with improvements in the airframe of the plane, Allen believed he could keep the plane aloft for hours, or until his own "fuel supply" of glucose and water was exhausted. The actual Channel crossing, impeded by head winds, required Allen to pedal for two hours 49 minutes. The distance covered was 22.26 miles; the average ground speed was 7.9 miles per hour.

The third success for the Betz-Prandtl theory came with the design of the propeller for the full-scale *Chrysalis*. Constructed between March and June, the biplane made more than 320 flights during the summer before it was finally dismantled in September. If *Chrysalis* had not had a theoretically correct propeller, matched to its airframe and power plant (.5 h.p.), I doubt that with its short wingspan and substantial weight (10 kilograms more than *Gossamer Albatross*) it could have flown at all.

Compared with the problems of the designer of airplane propellers the problems of the designer of ship propellers are far more constraining. The screw diameter is limited on the one hand by the maximum permissible draft of the laden ship and on the other by its waterline when the ship is lightly laden. The propeller must ingest a heavily nonuniform flow field, distorted by the contours of the stern and full of velocity defects produced by the hull boundary layer. In spite of a favorable multiblade geometry, the sharp limitation on screw diameter leads to excessive disk loading and to large induced losses. The solidity of the propeller, already large because of the heavy disk loading, is further increased by the limits imposed on blade angle of attack and camber by the necessity of limiting cavitation in the flow. Cavitation results when the fluid velocity near the blades is raised so high that the local pressure falls below the vapor pressure of water. In effect the water boils, suddenly creating cavities in the flow field. Cavitation is the principal source of marine-propeller noise.

Further difficult problems are presented by resonant frequencies of excitation that can be set up between the turning blade and the hull of the ship. Marine propellers are usually given curved blades with swept-back leading edges in order to make the excitation less sharp as a function of blade-rotation angle. For all these reasons the analysis of ship propellers must take into account the variation in inflow velocity over the entire blade surface and account for the roll-up of the vortex sheet, which takes place very quickly with heavily loaded propellers. The simple approximate distributions of inflow velocity and circulation of Betz, Prandtl

RUNNING MARINE PROPELLERS are studied in M.I.T.'s marine-hydrodynamics laboratory with the aid of high-speed photography by members of the Department of Ocean Engineering. In this photograph, which shows the screw from above, the white region at the tip of the middle blade is a large bubble of water vapor, evidence of abrupt cavitation caused by a nonuniform flow field that simulates the wake of a ship's hull. Low pressures created on the "suction," or front, surface of the blades induce instantaneous boiling of the nearby water. As the blade turns, the large bubble collapses and gives rise to a stream of smaller bubbles. Cyclic cavitation reduces propeller efficiency, causes vibration and generates a characteristic acoustic "signature." Cavitation is reduced by increasing blade area and by optimizing blade geometry.

and Goldstein are not applicable to ship screws; the designer must fall back on semiempirical numerical methods of analysis. In the end the efficiency of marine propellers is usually about 70 percent. Naval architects accept this as the best that can be done given the nature of the strong practical constraints.

At present considerable study is being given to propellers for airplanes intended to fly at 80 percent of the speed of sound, which corresponds to about 850 kilometers per hour (about 530 m.p.h.) at cruising altitudes of from 10.7 to 13.7 kilometers (35,000 to 45,000 feet). The blades of such propellers will also have rotational tip speeds of 80 percent of the speed of sound, and when the forward component of velocity is added, the tip speeds will reach 113 percent of the speed of sound. At such speeds propellers will need to be multiblade in order to achieve good lift-to-drag ratios and will have high solidity. Hence they will strongly resemble marine propellers. Efficiencies of 80 percent should be achievable. The high-speed propellers will be noisy in operation (for reasons quite different from those that make marine propellers noisy). The noise, however, will be chiefly "near field" noise, which can be made tolerable by improving the sound insulation of the fuselage. The "far field" noise may actually be less than that of existing turbofan engines. The fuel efficiency of a turboprop engine, given a suitable high-speed propeller, should be about 20 percent higher than the efficiency of a high-bypass turbofan engine with a "gas generator" (the combination of compressor, combustion chamber and power turbine) of the same size as the one that would be needed for the turboprop engine. The propeller of the turboprop engine would be about twice the diameter of the fan in the turbofan engine.

After more than a century of application in hydronautics and aeronautics the screw propeller is very much alive and turning; I expect it will survive for as long as man uses water and air for transportation.

TRANSONIC "PROPFAN" is one of several designs being studied by United Technologies Corporation and the National Aeronautics and Space Administration for potential service in turboprop aircraft capable of matching the speeds of turbofan airliners. If the new craft are successful, they should consume about 20 percent less fuel than present types do. The propfan is termed transonic because the speed of the blade tips will exceed the speed of sound by about 13 percent. The aircraft will fly at 80 percent of the speed of sound (about 530 m.p.h. at 35,000 feet). For operation in the transonic regime an aircraft propeller must approach the high solidity of marine propellers. The solidity of this propeller is 70 percent. Broad blades supply high lift-to-drag ratios needed for transonic operation. The curved blade planforms are chosen to reduce "near field" (cabin interior) noise caused by focusing of the rotating pressure field.

White Dwarf and Pedal-Powered Flight

BRYAN L. ALLEN

Bryan L. Allen first learned to fly in 1974 in a foot-launched hang glider. He was the pilot of the *Gossamer Condor* pedal-powered airplane that won the £50,000 Kremer Prize in 1977. The *Condor* is now on permanent display at the National Air and Space Museum in Washington, DC. In 1979, he piloted the *Gossamer Albatross* across the English Channel to win the Kremer Cross-Channel Prize of £100,000 for the group headed by Dr. Paul MacCready. In 1984, he set a speed record of 23.6 mph when piloting the *MacCready Bionic Bat*, winning the fourth installment of the Kremer World speed contest for stored-energy human-powered aircraft. He still holds two Federal Aeronautique International world records: distance and duration for Class BA-1 through BA-10 (nonrigid gas airships). All told, he has piloted eight different human-powered aircraft. Bryan is employed full-time by Telos Corporation, a software-services company, at the Jet Propulsion Laboratory in Pasadena, California. He possesses private pilot ratings for airplane single-engine land and gliders and is looking forward to some day having an airship rating.

The *White Dwarf* pedal-powered airship came about when Gallagher, a popular comedian, saw a model blimp in 1983 that had been built by a southern Californian named Bill Watson. He decided on the spot he needed one for his stage act. Gallagher, however, did not like the utilitarian style of Watson's demonstrator; he needed one that would make people laugh. A few weeks after their first meeting, Gallagher was on national TV guiding a Watson-built radio-controlled blimp with a bag shaped and colored like a watermelon (a Gallagher trademark), a rudder that resembled a child's kite turned sideways, and a little model Gallagher seated on an old-style bicycle mechanism vigorously pedaling away. "Whataya think?" he asked the audience. "Would you rather have one of these, or your car?" After mastering the art of radio-controlled flight, Gallagher would sometimes watch his blimp flying out over the heads of amazed crowds and think, "You know, that looks like fun." So he told Bill, "I want a blimp that will carry *me*." Bill did some calculations and consulted with other experts regarding this proposed blimp's structural and aerodynamic design; four months later, Gallagher was pedaling around the sky in his new toy, the *White Dwarf*.

The *Dwarf* was designed by Bill Watson who, among other accomplishments, was the chief construction engineer of the *Gossamer Albatross*, the pedal-powered airplane that flew across the English Channel in 1979. The airship's final design had undergone many compromises in order to make it small, maneuverable, strong, and quick to build. Gallagher had specified that he would like to fly indoors at some of his events; therefore straight-line speed, although not ignored, took a back seat to other goals set for this craft. This pedal-powered airship, Bill felt, had to pose the least-possible hazard to spectators; specifically, the propeller location, materials, and shape had to minimize the likelihood of anyone being struck. A larger, sharp-edged prop like the 13½-ft Kevlar and carbon-fiber creation on the *Gossamer Albatross* would have been about 10–15% more efficient than the 5½-ft foam and spruce prop used on the *Dwarf*, but would not have been as safe.

During the design phase, Bill discovered that Raven Industries in South Dakota made an envelope used to carry electronic payloads aloft that could be modified for use on a pedal-blimp. Using this bag meant the blimp would have more drag and go slower than with an optimized envelope shape, but the time saved allowed the *White Dwarf* to be built sooner than would have otherwise been possible.

The craft ended up having a Santos-Dumont turn-of-the-century flavor, with a gusseted and pop-riveted aluminum fuselage structure, recumbent seating position, and a large movable rudder. The rudder could move through a total

MUSCLE POWERED BLIMPS

Gallagher flies his pedal-powered airship, the White Dwarf, *near Camarillo, CA. (Photo © by Bryan L. Allen, 1984)*

arc of 150 deg and incorporated a large balance tab to capture and redirect most of the prop wash when turning. Altitude control was accomplished by vectored thrust, the propeller being tiltable through an arc of 100 deg (40 deg up, 60 deg down). The power train allowed pedaling in reverse, although at some loss of propeller efficiency. The propeller was geared through chains and sprockets to turn four revolutions for every one revolution of the pedals.

Bill and Charlie created a small radio-controlled model to get a feeling for the full-size blimp's control characteristics; the model suggested the *Dwarf* would be very maneuverable, but would be unstable in pitch at speeds above approximately 15 mph. This instability was not corrected for. With its exposed pilot, bare fuselage, and numerous external bracing cables, Bill did not expect it would be possible to fly the *Dwarf* faster than 10 or 11 mph.

The *Dwarf* took three people, each working a 40-hr week 2½ months to complete. The construction crew was Bill Watson, Charlie Sink, and myself. As with any prototype aircraft, there were many details that had to be worked out during construction. The control system, the seat, the fittings; all these and more were designed while we made them. We even built the trailer to haul the deflated airship around, assisted by Bill's brother Skip. Many samples of proposed components were built and ripped apart with a calibrated, hydraulic press test rig we made. By following the dictum, "Keep it simple and light, with safety first," we quickly found our solutions. One factor that distinguished us from many aircraft homebuilders was that we were interested in the end result rather than the building process. "Good enough, was a phrase heard many times during the construction of the *White Dwarf*. Yet we were careful to keep our standards high and always think of safety. We kept very much in mind the fact that the person we were constructing the blimp for earned as much in one night as any of us earned in a year and was a total novice at flying. At a total cost under $40,000, which included the trailer, tooling, a $15,000 prebuilt envelope, and generous wages paid to the three of us, the *Dwarf* still turned out to be one of the lower-cost prototype aircraft (that actually worked), pedal or otherwise of which I know.

As Gallagher has no pilot's license or formal flight training, we intended to operate the blimp under the FAA's Part 103 rules. These regulations specify a maximum empty weight of 254 lbs, a maximum fuel load of 5 gal, a maximum speed of 55 knots with maximum landing speed of 24 knots, and a limit of one occupant. Any aircraft meeting these simple rules can legally be flown by unlicensed opeartors in areas having light air traffic away from settlements. However, by calling the blimp an ultralight, we met with the hostility that some folks harbor against machines having "no license required." Many people in aviation feel *all* ultralight-category aircraft are inherently unsafe and unsound, feeling that ultralights and their operators threaten all other branches of aviation. Say "ultralight" to a lot of pilots and they are repulsed immediately; their reaction is as if someone suggested to devout fundamentalists that a Hell's Angels member be appointed the new minister of their church. We *wanted* to stay as far away from other aircraft as possible, but we needed a tall hangar to store the blimp in if we were to be able to leave it inflated between flight trials; therefore, we spent some time looking for airports run by people who at least had an open mind about ultralights. The *White Dwarf* required about 6000 ft^3 of helium at a cost betweem $800 and $1200; it just made sense that we would want to keep the craft inflated between flying sessions.

We were able to find a hangar and airfield at Camarillo, about 40 miles northwest of Los Angeles, that would allow us to operate on a very limited basis. The airport manager at first seemed to think we were a bunch of loonys, but ended up proposing we use a site under the base leg of the traffic pattern only yards from our hanger that was perfect for the tethered test flights we wanted to perform.

Tethered Flights

We found that operating this small airship (27-ft tall, 47-ft long, and just over 6000 ft^3 in volume) was easier than we expected. Turns with a radius under 50 ft could be accomplished, something we proved decisively by doing 180-deg turns inside our hangar two days before our first outdoor flights. True "hangar flying" is possible with such a small airship; with several similar blimps inside an enclosure like the Astrodome, you probably could have races!

Response from the vectored-thrust system was better than anticipated; when outdoor flights commenced, we found that moderate pedaling would net 200 ft per min (fpm) climb rates and 300 fpm descents. Contrary to our expectations and experience with other human-powered aircraft, it took only one person (with the pilot strapped into the *Dwarf's* seat controling the rudder) to ground handle the airship in winds as stong as 14 mph. And talk about easy to fly! After we put a mirror on the front pedal-mount post so each pilot could see the rudder position, not one of the over 40 people who have since flown the blimp has had any trouble adjusting

Bill Watson pilots the White Dwarf.

to the control system. "How hard do you have to pedal?" we answered with another question: "How fast do you want to go?" Pedal lazily and you would eventually top out at 4 or 5 mph; pedal like a bike racer sprinting for a gold medal and you would zip around (relatively speaking . . .) at 11 or 12 mph; 7-8 mph seemed to be a sustainable cruise speed.

Spreading the Fun

We let quite a few people fly the *White Dwarf* once we had completed our flight testing. I used to work for an ultralight manufacturer and soloed quite a few people in fixed-wing ultralights. I never got away from suffering sweaty-palmed stomach-knotting tension when watching students make their first powered solo. However, I was *never* tense soloing a person in the *White Dwarf*, even when, in one case, the student had never been aloft in *any* flying machine and was only given 30 seconds of verbal instruction. When we would allow people to fly the blimp for the first time, we would dangle a "guide rope" over the side that had five 2-lb weights spaced evenly along its 25-ft length. As the guide rope contacted the ground, the weights no longer pulled down on the blimp, giving in an excess of lift that made it rebound from the ground. We found it nearly impossible to make bad landings with this "landing gear." Training radios? Didn't need them; we just shouted up at the students. Student going the wrong way? We would give a tug on the 250-ft Dacron-sheathed Kevlar tether and he would come back whether he wanted to or not. This ultralight airship turned out to be an anxious ultralight flight instructor's dream come true.

Off-Tether Flights

After flying the *White Dwarf* on tether for several months, Bill and I decided to attempt an off-tether cross-country flight. Both of us have experience flying radio-controlled gliders, ultralights, hang gliders, and pedal-powered airplanes such as the *Gossamer Albatross*. This experience sensitized us to the very small-scale weather effects we expected would bother the *Dwarf*. Our goal was to go three times as far and stay up three times as long as was done with the *Gossamer Albatross* on its 22.3-mile, 2.8-hr English Channel flight in 1979. The area we chose for off-tether flight operations was along the northern edge of the Salton Sea, our base being the large hangar at La Quinta Flying Service in Thermal, California. Aerial scouting from a Cessna 172 led us to believe that a flight southeast from Thermal along the eastern edge of the Salton Sea looked like the best prospect for a long flight in California during the wintertime, although we knew care would have to be taken to avoid military airspace over and around the water. We felt that flying over water during midday would allow us to avoid atmospheric convection that otherwise might compromise control and safety.

We drove the *Dwarf* in its trailer to Thermal in early February of 1985. After dropping off the trailer, we went to the local welding-supply shop where our preordered helium was waiting. Our 1-ton pickup staggered to the airport under the 3500-lb load of 25 helium tanks, mute testimony to why we do not recompress the helium from the *White Dwarf* when it is deflated; the additional expense of transporting nearly 2 tons of tanks plus the cost of a high-capacity compressor unit would be much more than just letting the helium go. Capturing the helium in a lower-pressure receptacle does seem feasible; we have not pursued this, but I am sure we would if we (rather than Gallagher) owned the blimp.

After inflating and rigging, we started to learn about flying off tether. One week had been set aside for testing and practice before I would attempt a record flight. The first morning Bill flew, then I did. We were both very cautious, staying within 50 ft of the ground and no more than a few hundred feet from the ground crew. We were concerned about sudden weather changes that might overpower us and waft us away. We suspected the strong thermals could lift us too high, too fast; the *Dwarf*, to be as small as possible, does not have a ballonet to accommodate envelope pressure changes, but only a spring-loaded manually operated valve for pressure regulation. We did not know how fast the bag pressure would rise as we ascended, But only that Raven Industries, the bag's manufacturer, expected the envelope would burst when pressurized to 17 in. of water. We had a lot of questions, but there was no one we could ask who knew anything about free flying a pedal-powered airship.

Even the first tentative free flights felt different from tethered flight. Both Bill and I experienced a tangible awareness of freedom that we had not noticed previously when flying the *Dwarf*. Any tendency to exult in this new-found freedom was tempered by the realization that we now had the "freedom" to blunder into any nearby powerline, church steeple, or other obstacle.

At around 10:30 a.m., we found thermal activity would increase to the point where flying was uncomfortable. It was necessary during morning flights to vent a small amount of helium every half hour or so (due mostly to the solar heating of the bag) to keep the envelope pressure within the conservative (we hoped) maximum pressure limit of 2 in. of water on which we had decided. Because we would normally fly the blimp only 1 or 2 lbs "heavy," any increase in envelope pressure would stretch out the bag a bit and make the blimp overly buoyant. Since we had no ballonet, the only way to deal with being overly bouyant was to vent helium, land, and add more ballast. This was not as easily done as said. Most pedal-powered aircraft are difficult to keep *up*; we sometimes found it difficult to keep the *White Dwarf* *down*.

Late-afternoon flying had *quite* a different character. The lessening solar input worked to make the blimp heavy, which we joyfully dealt with by releasing water ballast in flight; joyfully, because water costs nothing, plus this gave the pilot myriad opportunities to rain down on unsuspecting targets.

On the following day, we got a bit more adventuresome and the air a bit more unstable. After some gentle early-morning flights by Bill, myself, and our fellow builder, Charlie Sink (who decided that 20 ft was plenty high enough, even though he had been higher on tether), the convection started to be in excess of 200 fpm. In these conditions, I decided to stay within just a few feet of the ground to hopefully avoid the strongest lift. This tactic worked for a while; then I flew into some gradual lift west of the airfield that lifted me 100 ft or so. I decided that I would try putting in a full down vector on the prop; I knew with vigorous pedaling in calm air this would yield a descent rate of up to 300 fpm. Checking the variometer, I found I was still ascending at about 100 fpm while working fairly hard. "No problem," I thought to myself, "I'll just turn

> Bill cavorted about a water tower and delighted in the powerful but dream-like feeling, when climbing under power, of walking up an invisible staircase into the sky.

180 deg and fly out of the lift; I haven't been in it very long." I turned to the east and accelerated to about 9 mph indicated air speed, confident that I would be out of the lift in no time. After a minute, I noticed that I had made no progress over ground and was, in fact, still over the tree where I first noticed the lift. "Must be a breeze up here," I thought, "I'll turn and head downwind to the west." By now I was about 800 ft above ground level (AGL). After changing course again, I was still over the same spot! I vented helium for about 20 seconds, enough I knew to make me descend at about 400 fpm when in calm air, and turned from side to side in search of sinking air. Now I was in the middle of the airport traffic pattern, fearing a collision, my parachute back on the ground, and still ascending. Finally, I was gently buffeted by small puffs of turbulence, the variometer indicated my climb had ceased, and I noticed I was actually tracking away from that lone tree now far below. I vented helium for another 20 seconds, noting my altitude as 1200-ft AGL, and slowly started down. Although I had just set an unofficial altitude record for human-powered aircraft, I was disgusted and a bit frightened by the experience of being helplessly wafted upward. Once I was in the thermal, it was as if *every* direction was upwind. It took me another 30 minutes of searching for sinking air before I got close enough to the ground crew so that they could grab the 200-ft long landing line I tossed overboard. That evening, several of us talked over my flight and concluded that we had never fully understood the true structure of thermals. We felt that there must be far more air being pulled into the sides of the thermals than textbook explanations indicated. We later concluded that the effect is most likely an aerodynamic one, similar to what occurs when a beach ball is suspended by a fast-moving column of air above a blower. We could see only two remedies for dealing with thermals: one, fly faster (not really possible with the *Dwarf*); or two, avoid thermals.

A subsequent morning was overcast and windless. Both Bill and I took turns roaming far and wide in the blimp, chasing hawks, skimming over palm orchards, waving at passers-by. The convection that day never seemed to develop. Bill cavorted about a water tower and delighted in the powerful but dreamlike feeling, when climbing under power, of walking up an invisible staircase into the sky. I flew a triangular course north of the airfield that was 8-10 miles in length, getting a feeling for steady-state pedaling. During this flight, I startled a farmhand working 50 ft up in a palm tree when I flew 10 ft over his head and rang the bicycle bell the *Dwarf* carried. From the look on his face, he must have thought I was a hostile visitor from another planet.

The Record Attempt

We assembled a small group of volunteers and our appointed National Aeronautical Association (NAA) observer, Dan Glick, at Thermal airport on the morning of February 12th. Bill and I carefully checked the blimp. Its installed instruments were an airspeed indicator, electronic sailplane variometer, bag pressure gage, ambient-air thermometer, altimeter, and, on my wrist, a calculator watch. I had rigged a drinking tube from the forward ballast tank (which held about 30 lb of water) to provide me with liquid and also strapped to the fuselage two bicycle water bottles. For "fuel," we attached a container that held several sandwiches and pieces of fruit.

Takeoff was at 7:23 a.m. Initially, there was light tailwind that whisked me south toward the town of Mecca. I kept in touch with the Flight Service Station at Thermal using a small hand-held aviation transceiver taped to one of the fuselage tubes. I could talk to my chase crew (plus passing truckers and vacationers) with the CB radio we had affixed to the *Dwarf*. With all this gear, the cockpit ambiance was far removed from the spartan feel of the *Gossamer Albatross* that I flew across the English Channel in 1979. That craft carried only 2 liters of water, no food, one very stripped down transceiver, a quarter-ounce bimetallic thermometer, and two ultralight custom-built flight instruments (airspeed and altitude) that had just enough batteries to function for a 2-hr flight.

As I approached the northern edge of the Salton Sea, a headwind sprang up. Two hundred ft below me, a jogger kept pace, even though

> **Building a human-powered aircraft can be an appropriately sized task for a student group. The return on investment of time and funds can be very good if the goal chosen is easily defined and achievable.**

A record in the making.

I had the advantage of being able to cut diagonally across the field and orchards. My ground speed dwindled to 3 mph; things did not look good. After an hour of very slow progress, I broke through to a southerly flow of air that, with my steady pedaling, allowed me to "clip" along at 14 mph ground speed. Flying along the sea's eastern edge, I found I had to maintain 200-300 ft in elevation and closely parallel to the shore in order to keep from being either swept west over the water or east over land. It was a delicate balance, with contrary air currents just above and below me. I occasionally valved helium to stay in equilibrium, while pedaling and eating.

By noon, I was over the small town of Bombay Beach. Here, the shore bent almost due east. I decided, after checking my banana supply, to cut southeast across the bay and shorten by flight path a bit. As my chase crew had no boat, this meant I was completely on my own. I kept a steady pace, watching my shadow on the water below to estimate my progress. As the afternoon went by, thermals bumped me around some, even over the water, and headwinds again slowed my progress to walking speed. By 2 p.m., I was back over land, the headwinds were decreasing, and the thermals were lessening in strength. A few agricultural planes flew alongside to check me out, thankfully not getting too close. I pedaled onward, starting to tire a bit. By 3:30, the town of Brawley was just ahead. We decided to terminate the flight at Brawley airport; I floated about lazily for 45 minutes while we waited for the local TV station to show up and film the landing. After 8.8 hr I landed, we deflated the blimp, answered questions from the spectators, went into town to eat dinner, and then drove back the 70-plus miles to Thermal. The official straight-line distance between the takeoff and landing points was just over 58 miles. The NAA and Federation Aeronautique Internationale (FAI) officially recognized my flight as establishing two new world records in the Gas-Airship category (Class BA-1-10).

I was a bit sore the next day, noticing that it was more comfortable to walk up stairs one by one rather than my normal two at once, but suffered no other ill effects. For a

Table 1 Comparative achievements, performance, and other qualities of record-setting human-powered aircraft

Aircraft	Airplanes				Airships	
	Gossamer Condor	Gossamer Albatross	Michelob Light Eagle	Daedalus 88	White Dwarf	Low-drag pedal airship
Date of record flight	August 1977	June 1979	January 1987	April 1988	February 1985	1989?
Distance flown	1.4 miles	22.3 miles	37.2 miles	74 miles	58.1 miles	150 miles
Duration	0:08 h	2:49 h	2:19 h	3:54 h	8:50 h	10 plus h
Wind during record flight	Calm	4-mph head (average)	0 mph (average)	4-mph tail (average)	0 mph	—
Airspeed (cruise)	10 mph	11.5 mph	16 mph	14.5 mph	8 mph	15 mph
Weight (empty)	70 lb	70 lb	90 lb	72 lb	130 lb	120 lb
Power required (estimated)	.36–.40 hp	.22–.25 hp	.22–.25 hp	.20–.23 hp	.20–.23 hp	.20–.23 hp
Structural	Aluminum	Carbon fiber	Carbon fiber	Exotic carbon	Aluminum	Carbon fiber
Quality of Construction	Acceptable	Very good	Excellent	Magnificant	Very good	Very good – excellent
Work hours to build	1500	1400	12,000	15,000	1200	1500

comparison, after the cross-channel flight of the *Albatross*, I had sporadic leg cramps and spells of dizziness as long as two weeks afterward.

Why have blimps not been used in all previous human-powered flight attempts? I believe this happened because all prizes for pedal flight *specifically disallowed* the use of any lifting gases. These rules had as their result many years of frustration, followed by the eventual technical triumphs of the *Gossamer Condor, Gossamer Albatross,* and most recently, *Daedalus 88.* Coincidentally, the British group that wrote the rules for the Kremer Prizes that spurred on pedal-plane designers worldwide is named the "Man-Powered *Aircraft* Committee" (my emphasis). Maybe, it is time for a "Human-Powered *Flight* Committee" to come to the fore. The journey from Thermal to Brawley in the *White Dwarf* was more challenging to my navigational and flight-judgment skills than my cross-channel flight in the *Albatross,* which mostly taxed my powers of concentration and athletic skills. Navigating through the air is always a challenge, no matter what type of machine you choose to employ. The *White Dwarf* allowed me (and would allow any other reasonably fit person) to set human-powered flight distance and duration records equal or superior to those possible with *Daedalus 88,* plus it is more fun to fly and took less than 10% as much time (and, probably, money) to build (see Table 1). The *Dwarf* is also much stronger, with a 5-g ultimate load capability, which makes flying hundreds or thousands of feet above the ground feasible. All cross-country-capable pedal airplanes have been extremely fragile. To prove this point, the *Albatross* suffered a broken wing moments after a perfect landing on French soil and the *Daedalus 88* broke up just before landing on Santorini. Pedal blimps can carry more load and be far sturdier than pedal-powered airplanes without suffering major performance penalties. Pedal blimps have a future because they offer performance, cost, and safety advantages over pedal planes.

In April 1977, Greg Miller left the *Condor* project and was replaced as principal pilot by Bryan Allen. Here Bryan, on the left, shakes hands with Greg after a streamlined bicycle race in 1979. Behind them is Bill Watson, a member of the *Gossamer Albatross* team.

Under the direction of Dr. Joseph Mastropaolo, Allen began a strenuous training program to improve his ability to fly the *Gossamer Condor*. Here he works out on an ergometer.

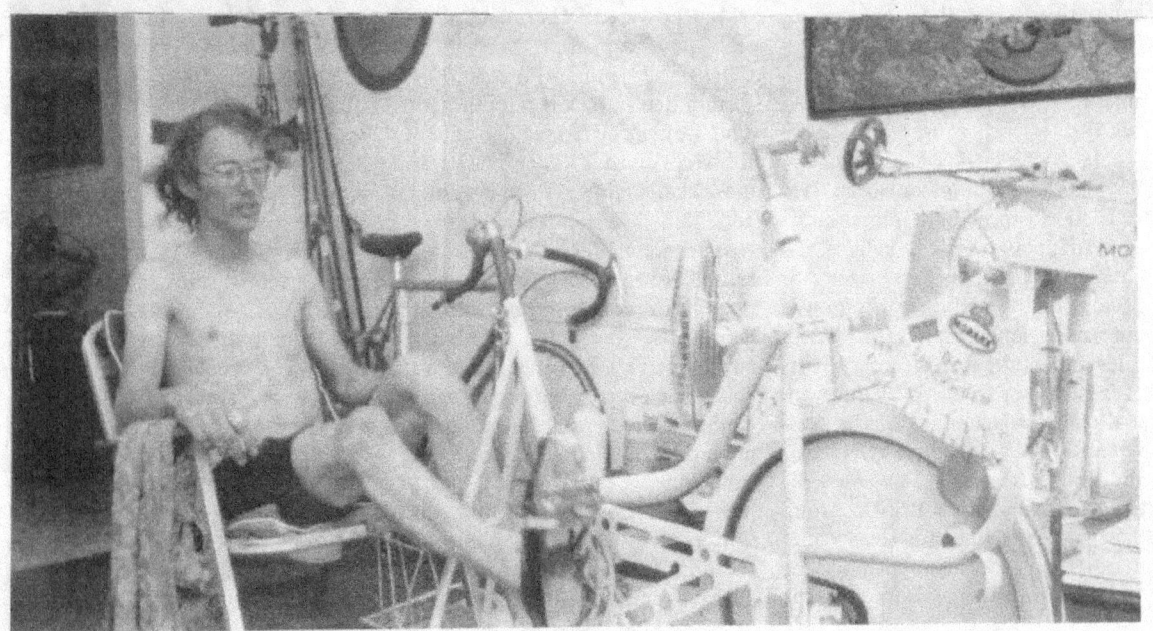

A reprint from ***MODEL AVIATION*** May 1985 p80
(White Dwarf)

MAN POWERED BLIMP
By Bill Warner

The comedian, Gallagher, wanted a non-polluting little cloud to encourage people to look upward and think about what they were doing to the air, so he provided the incentives for Bill Watson to develop man-powered airship—the subject of this article. BILL WARNER

CAMARILLO, CA is the site of the State Mental Hospital. Suppose one of the passengers on a city bus suddenly began waving to the driver about seeing a man pedaling a little white cloud around over the airport. You know what the reaction would be: "Sure, sure . . . and the guy is all green with yellow polka dots, right?" Suppose the driver took the trouble to look for himself; he might consider checking himself into the hospital for observation! For that maker, whenever I tell anyone that I flew a pedal-powered blimp all by myself, I get somewhat the same reaction. If I didn't have the same reaction. If I didn't have the photos to prove it, I might wonder if it was all in my imagination!

The genius behind the "little white cloud" is Bill Watson of Van Nuys, CA. I've known him since his school days back in the Sixties. He used to come to our Sunday Indoor Scale meets at Sepulveda Junior High with homemade blimps made from dry cleaners' bags that were heat sealed with a soldering iron (applied through a layer of newspaper to keep it from sticking). They were rubber band-powered and filled with hydrogen that Bill had manufactured with a complicated little contraption he had made. Playing with hydrogen is dangerous, but we didn't have cyanoacrylates and boron filaments to tempt fate with in those days, so we needed something to keep it all interesting. His blimps were always the hit of the meet, and they puttered about in circles for minutes on end.

Bill went with me to the Museum of Science and Industry when we started the Aeroplanes and Flying Machines class in Los Angeles. Subsequently, hundreds of youngsters in Los Angeles made their own blimps and flew them successfully (they *all* fly). we made some from dry cleaners' bag material, but later we went to the Zeppelin balloon for the envelope, putting the "gondola" made from material in a Styrofoam egg carton directly on the balloon with vinyl electrician's tape. The motor stick was balsa, and the "businselage" of the Golden Football supply was a balsa stick, an aluminum tube bearing, and a prop the kids made from a stack of balsa laminations that were fanned out, glued, and sanded. In a pinch we'd use a heavier plastic prop. The fins were cut from construction paper stuck on with Magic Mending Tape. About a 2-ft. loop of 1/16-in. rubber would give flights lasting 2 min. in a room with a 12-ft. ceiling. The blimps bounce off the walls!

A few years later, Bill Watson was making Radio Control blimps of some sophistication that was a sight to behold at the Pasadena IM Show. His dozen-foot-long Golden Football used 48 cubic feet of helium (a welcome relief from the flammable hydrogen) to provide the lift. helium in its aluminized Mylar bag. The lifts a little less than hydrogen, but it is a whole lot safer. Too bad that the Hindenburg, designed with bags for helium surrounding hydrogen cores, wasn't able to get it; as a last resort, they used hydrogen all the way (or almost all the way) and had the disastrous consequences which we know so well. Helium lifts a little less than 1 oz. per cu.ft. which is quite a bit more than you can do with hot air.

The "fuselage" of the Golden Football was suspended about a foot beneath the shining eight-gore DuPont Kapton (a Kevlar film) bag. The electric motor, an Astro-02, was rigged for rudder-aided directional control. Longitudinal control and stability came from a fixed stabilizer in the rear and a movable one (like the diving planes on a submarine) on the front. It carried 5 oz.

MUSCLE POWERED BLIMPS

A reprint from *MODEL AVIATION* May 1985 p80
(White Dwarf)

of water ballast. The motor could be controlled to give reverse in flight. (Try that one on your winged RC airplane!) A nice feature of the design was a release mechanism for dropping things (like small hang gliders) from altitude.

Bill's most recent RC blimp drops things too, but it is much smaller and more practical for small gyms. It carries 17 cu.ft. of helium in its aluminized Mylar bag. The covering is Saran-coated, and it seals easily with heat. Bill says he doesn't know where you can get it, though. His high-tech materials came to him via the team that built the famous Gossamer Albatross and Gossamer Condor. Bill was a part of both projects and knows more about super-light construction than almost anyone I know.

The entire aluminized blimp is 6.5 ft. long, but it has only 3 oz. of material in the gas bag. It is a two-gore design—simple to make but having lots of wrinkles (not critical when you only fly at 3 mph). The three-channel radio operates motor speed (the power motor is like the type used in servos), thrust vectoring up and down, and rudder. The rudder is tied to the gimbaled motor mount; applying rudder moves the motor thrust sideways to give turns on the proverbial dime. The up and down control provides a range of thrust from straight up to straight down. It is so maneuverable that it is often pressed into service to retrieve Peanut Scale and microfilm-covered ships from overhead lights at the Paul Revere Junior High School Indoor meets.

Long interested in human power, Bill not only was helping build the Paul MacCready Gossamers, but also hang gliders, bicycles for speed record attempts at the Ontario Raceway, and shells for racing bicycles. His inventiveness eventually brought him in contact not only with the top people in human-powered vehicles, but also got him building unusual stuff for the movies. That was his Radio Control pterodactyl in the movie, *War Games*.

One contact that changed his life was meeting the popular comedian, Gallagher. Gallagher is not your run-of-the-mill comic, but a highly-imaginative and creative philosopher whose love for humanity and nature drives him into some pretty zany demonstrations. The idea for a non-polluting little cloud which could be pedaled about (and in the process provide a vehicle for getting people to look skyward and think about what they were doing to the air) was born. Bill was given carte blanche to make one, though there were none to copy—the golden age of the blimp having peaked some 75 years earlier.

Drawing on his model experience, Bill did the basic design work—kicking around his ideas with Blaine Rawdon and Bart Hibbs, who provided invaluable assistance in stress analysis, configuration, and computer-designing the prop (a la Larrabee, for those of you who know about computer-designed props). An order was placed with the Raven firm, which makes all types of hot-air and gas-filled envelopes) for a tear-drop-shaped envelope of ripstop nylon with a urethane coating to be made with 16 gores. The more gores, the less wrinkles; the fewer wrinkles, the less drag. The finished bag is 47.5 ft. long and weighs in
he neighborhood of 60 lb.

Brian Allen (pilot and power plant of the great Gossamer flights, and incidentally the winner of more prize money than any pilot in history in so doing) and Charlie Sink, one of the graduates of the Aeroplanes and Flying Machines class were called on to work with Bill on the fuselage. Built of 2024 T-3 aluminum alloy tubing and 7075-0 gussets, the fuselage is tremendously strong. The gussets were pressure-formed in dies that Bill made, and pop-rivets fasten them to the tubes in their Warren-truss configuration. The prop has a spruce spar and main blades made of Styrofoam. The prop is driven from a pedal crank and standard bicycle chain connected to two "plastic-cable" chains of the sort used on the Gossamers.

MUSCLE POWERED BLIMPS

A reprint from *MODEL AVIATION* May 1985 p80
(White Dwarf)

A lever on the pilot's left controls the vectoring of the propeller thrust up or down—not as much as on the RC blimp, but enough to make the difference between a climb or descent. A small crank on the pilot's right controls the rudder position, which is monitored on a small rear-view mirror. The rudder, itself, features Styrofoam rib cores and spruce cap strips. It is double-covered with thin Mylar film. Total fuselage weight is 60 lb.

Ballast tanks, fore and aft, are of fiberglass and can be emptied of water in about a minute by the pilot. Easily-attached lead weights compensate for carrying different loads when there is a wide variation in pilot weights. An "equilibrator," consisting of several lead weights on a rope, contacting the ground progressively, is normally carried in a bag at the side of the fuselage.

Until a new pilot has a few flights under his belt, it is hung down under the airship, and a too-rapid descent becomes a docile touch-down as each of the lengths of the rope lightens the craft. This is a much-appreciated feature by pilots of free-flying gas balloons, that of retrievable ballast.

A model airplane motor and prop adorn the front of the fuselage. The belt-drive Astro 40 with an 18-6 prop can be turned right and left by the pilot. In theory, it is used to help with sharp corners or to buck headwinds. In practice, it has little effect, and it may be removed in the future.

Instrumentation includes an audible variometer (the buzz pitches higher as you go up), an airspeed indicator, a bag pressure indicator, and an FM radio for ground communication. Actually, yelling between the craft and the ground makes the radio unnecessary. The airship makes no noise to compete with the yell!

The attachment cords between the fuselage and envelope are Kevlar surrounded by Dacron for ultraviolet protection and abrasion resistance. Reinforced round patches glued to the gas bag distribute the load. In reality, there is a large safety factor in having many more than the minimum number of attachment cords. Even the seat belt has a backup should one come unbuckled.

Inflating the bag takes a lot of gas. At 16 ft. diameter, the bag requires over 20 tanks of helium (about 6,000 cu.ft.). If the price of the helium does not bother you, the $13,000 for just the envelope may start you thinking in terms of a smaller, model blimp! Even Bill is now building a smaller, living-room version for RC Electric.

We arrived at the Camarillo Airport for our first glimpse of the magnificent airship early in the morning. You can't fly in the wind, and the dead-calm from about 7 a.m. to 10 a.m. is about all you can do for one day. (In the Houston Astrodome, you can fly anytime!) With a top speed of around 9 mph and a cruse of 7 mph, you are at the mercy of the elements.

The airship, temporarily nicknamed the White Dwarf pending a contest to officially name t, was walked out to the flying spot which was a few hundred feet wide and bounded with hangars and airplanes. The tether, a rope of some 50 to 70 ft., was attached just in case a gust or miscalculation should imperil the ship. It was left slack, and the blimp would fly "on its own."

After a couple of shakedown flights, Bill Watson graciously consented to let me have a go. I will admit to being absolutely fearless right up to the time I was about 40 ft. off the ground, when I discovered that any incipient acrophobia in a person was going to make its presence felt in this rig!

Before liftoff, the bag pressure (about two to four hundredths of a pound per square inch) is checked. There is an automatic valve to let out excess pressure if it gets too great, and the pilot has access to a normal (but seldom used) pressure release valve. No problem. Bill and Brian

MUSCLE POWERED BLIMPS

A reprint from *MODEL AVIATION* May 1985 p80
(White Dwarf)

adjust the water levels in the ballast tanks to make the ship about 1 lb. heavier than its hovering weight (zero).

After weighing off, it's time to pedal with the prop pointing upward to climb. The plane of the lever approximates approximates the plane of prop rotations so you know how much vectoring you are giving it. It takes a while for the pedaling to take effect. The ship responds to thrust and rudder changes with what seems to be about a 15-sec. delay. Still, who cares?

I stopped pedaling and took stock of what to do next. The ship just hung there. Fantastic! Well, let's center the rudder and go! If you pedal harder, you go faster. If you pedal backwards, you slowly stop and then Begin to back up. The blades on the prop are adjustable on the ground, and the new, lower-pitch setting used this day, says Bill, gives better speed in both directions. What a view from up here! It may be slow, but it's as sure as sitting in front of your TV set. The two seat belts felt good, though, when balancing on the narrow fiberglass, semi-recumbent seat.

Rudder hard over, moving well—about 6 mph. I think I could get to like this! Quiet except for a swish-swish-swish of the prop turning at 330 rpm. After a few circuits of the course, I was ready to come down. Pedaling tends to tire the novice easily—and as I mentioned earlier, I found that I was not really too crazy about heights (though flying has never bothered me in the least before). In any case, it was a thrill never to be forgotten.

As the morning progressed, a wide variety of experienced and novice pilots took the blimp for a spin. Gallagher, Bill, Charlie, Brian, and a couple of the other old hands certainly had the rig well in hand and could make it go exactly where they wanted. Newcomers (like my wife, Phyllis, Bill's 12-year-old nephew, George, and a lady who ran a kindergarten) took their first flights, mastered the ship, and loved every minute of it.

The Peripatetic Raindrop (my name for it) is registered as an ultralight aircraft. At a flying weight of 1 lb., it is lightest of the ultralights in the world! According to Brian Allen, the ship is ". . . ludicrously easy to train someone in, and it is four times as strong as anything else." Besides that, says Brian, how many other human-powered craft can be flown indoors?

The possibilities of model and powered blimps are many. Gallagher has used an RC model which looked like a watermelon on his TV show. Peck-Polymers has a quite inexpensive blimp, being sold more and more for promotional stunts—and several countries are interested in them for military purposes. Aside from its eye-catching advantages in advertising, the human-powered one might be set up to give rides, maybe selling a Polaroid shot or videotape to the customer afterwards. We might even have a human-powered blimp race event at the next Olympics!

If you can't afford one of the big ones though, remember how it all got started—with rubber-powered models. Secrets? None, really. Just add weight until your model just hovers, and then add just a tad to bring it down when the power runs out. If your prop is lower than the gas bag, the airship will climb naturally, as the thrust is continually pushing up the nose into a positive angle of attack. Even the Goodyear blimp works that way. The vectored thrust is icing on the cake if you want to go RC. Happy flying, and stay out of the cactus!

(Editor: The March 1981 **Model Aviation** contained two excellent articles on RC model airships: the Slow Year Blimp by Lou Bruhn and the Sumaner Breeze Balloon by Luther Hux.)

Article reproduced in its entirety for the
Academy of Model Aeronautics

MUSCLE POWERED BLIMPS

The Flight of the Brazilian Dirigible *CALOI*

by Zilda Camacho (author & photographer)

A dirigible developed & constructed with Brazilian technology made its first flight at Campo de Marte on 14 September. It was piloted by Marco Antonio Martin Migiano who proudly displayed the new technology of the *NEPTUNE Group*. It made its second flight on 24 September in front of the Ipirapura Obelisk in Sao Paulo, in full view of the passing "Tour de Brazil" bicycle ragatta, sponsored by the "Caloi" Bicycle Company (a $40,000 investor).

The second flight that lasted an hour and a half, was interrupted for safety reasons, as the earlier winds of 4 meters per second, increased to 30 meters per second. The project was developed and built by the Neptune Group at the Naca Company facilities in Sao Paulo and coordinated by its engineer/president Karl Gruschka.

The experimental pedal powered dirigible idea was first proposed by Caloi to be accomplished by another private group. Besides Caloi, the Air Products Company agreed to help on the project, giving a special price on the helium gas of about $5,000 USD[1].

The Neptune Group dirigible, called BOBIC (ed. bio bubble study) will be shown at public exhibitions around Brazil. A definite program schedule according to the builders.

Like all Neptune Group projects, this one started with a "work force" in place. All of the members had one interest in common; the goal of developing and building a flyable dirigible. Like the Bobic, for example,

Undertaking this idea in Brazil was complicated, as there is little incentive or interest in research or in developing technology in general. The Neptune Group is a non-profit group, that depends on private investment to realize these projects.

In 1986, Nelson Barbosa Corea who was acquainted with the groups potential, contacted the Neptune Group. He suggested that we start a separate company to sell the groups development projects, and the NACA company -- Neptune Commercial Activities & Assistance was created.

The Neptune Group had originally contacted Caloi in 1985 as a tentative sponsor. With the creation of NACA In 1987, a new contract was made, and this time they easily agreed to become a BOBIC sponsor.

Once Caloi was a sponsor, Neptune had no problem interesting them in a pedal-dirigible. And from that moment the data that Neptune had collected was utilized and new data developed. During the 15 months of project development much research and experiments were undertaken on the materials necessary for building the dirigible. For the ideal fabric, many weeks were spent experimenting, until they decided on polyurethane coated polyester, never before used in Brazil. It proved ideal, light, water resistant, impervious to the gas and still elastic on the aircraft.

With the research started, much technical data had to be determined: weight, volume, speed of flight, propeller diameter and speed, transmission design, etc. Also testing, strength of structures (assuming 4G loads), the permeability of the fabric, and the most efficient pilot position.

CALOI PEDAL-POWERED AIRSHIP cont.

Training plans and procedures for flying in different wind speeds, radius of rotation; coordination of tractor prop inclination; pressure control; pitch control, ballast extremes; weight & balance control. These items and procedures had to be firm estimates for the safety of the pilot and dirigible.

All of this data was necessary because the (envelope volume) in normal conditions has a specific float altitude. Higher or lighter requires more air (some grams) at the altitude that one is supposed to fly. These are safety measures, because when the pilot cuts the power, the dirigible automatically descends to the ground at low speed, . On takeoff with no elevator, the tractor (prop transmission) is tilted upwards as the pilot pedals so the dirigible climbs slowly. When the altitude is convenient for cruise flight, the pilot must balance the dirigible to maintain a convenient float altitude. The pilot needs to maintain equilibrium to float at that chosen altitude. When descending, the gas contracts, and is compensated for by increasing the air volume inside the envelope "ballonet". This can be done by forcing air into the ballonet. In case the pilot wants to descend quickly vector the prop slightly down and pedal until the dirigible is on the ground.

In stronger winds it is possible for the pilot to maneuver the landing to an unassisted tie-up. For this, he can throw out a three point steel anchor with an 80 meter rope, that will attach itself to the ground. The wind is is a very critical factor in the operation of this type of dirigible because of the low drive speed. Bobic has a maximum design speed of 15 kph, but only a realistic (pedal) power of 6 kph. But its big safety feature is that the low pressure of the envelope allows, that should any leak or accidental hole appear, the blimp simply floats slowly to the ground.

According to calculations by engineer Karl Gruschka, inventor and director of the project reported that during one operation, a leak in the envelope caused by a 4 inch hole takes about 4 hours to deflate, but that depends greatly on the position of the hole and any weight on it.

Variations of pressure, caused by heat (called "superpressure"), are controlled by keeping the gas pressure low before take-off. If the pressure increases excessively, as can be precalculated, a safety valve automatically releases the excess gas.

To fly the aircraft, the pilot does not need a lot of technical aerodynamic knowledge, or need to be an airplane pilot. It is enough to only have a little instruction on the ground and a few minutes on a tether, until he learns to control of the aircraft.

Other dirigibles are being planned for different uses, but many new sponsors will be required. The apparent demand will require engines for commercial and sports events, pictures & films, police traffic reporting, ecology research, aerial mapping, tourist flights, and advertising publicity. Many benefits of dirigibles in Brazil were mentioned in the July 1985 issue #537.

The public display was possible thanks to the Civil Aviation Department (DAC) who solicited the use of the facilities at the Aeronautical Park in Sao Paulo (PAMA) for the final assembly of Bobic.

The Minister of Aeronautics has stated that experimentation is directly responsible for the technological development of aviation. The Director of PAMA is commended for his cooperation in providing the space and technical equipment for working on the dirigible final assembly.

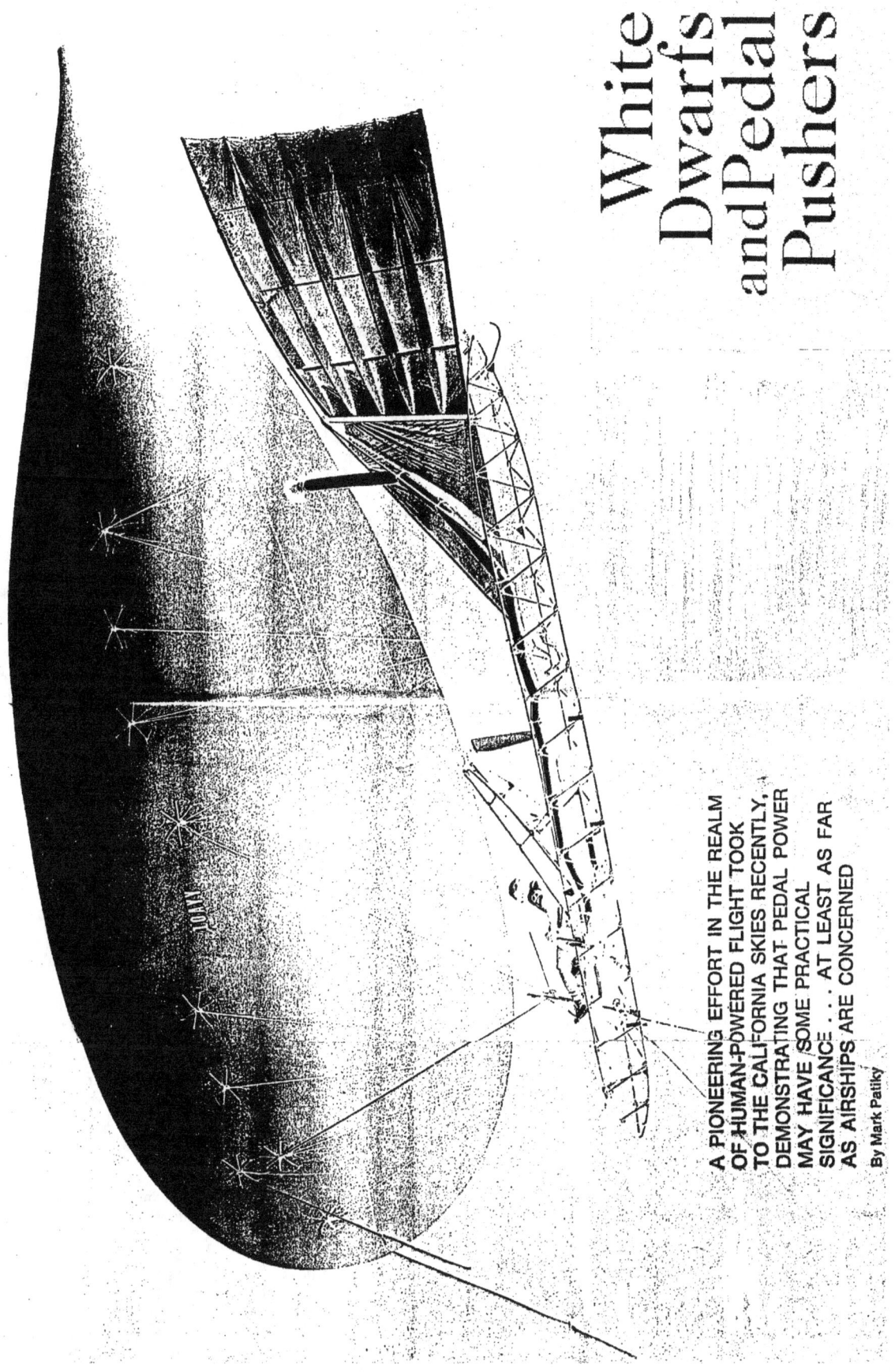

White Dwarfs and Pedal Pushers

A PIONEERING EFFORT IN THE REALM OF HUMAN-POWERED FLIGHT TOOK TO THE CALIFORNIA SKIES RECENTLY, DEMONSTRATING THAT PEDAL POWER MAY HAVE SOME PRACTICAL SIGNIFICANCE... AT LEAST AS FAR AS AIRSHIPS ARE CONCERNED

By Mark Patiky

MUSCLE POWERED BLIMPS

It hung motionless in the sky— a great white object forty-eight feet long. At first it hovered like some alien spacecraft and then slowly moved closer emitting earth-like sounds. A honk, honk from an old-fashioned car horn and then the ring of a bicycle bell. As it came closer it was obvious that the gondola carried not a space cadet but a fast pedaling humanoid in the guise of one Bryan Allen. Indeed, it really was a man-powered airship . . . a blimp and an Ultralight no less!

Test pilot Bryan Allen has had more than a modicum of experience with human-powered aircraft; to date he's logged more time in this pursuit than anyone else in the world. He pedaled his way to victory with the Gossamer Condor, winning the renowned Kremer prize for the first human-powered aircraft flight around a prescribed course. Then, in 1979, Allen scored another human-powered victory as the first to pedal across the English Channel in the Gossamer Albatross. After a three-year stint in the Ultralight manufacturing business, Allen is back pedaling again in another first for mankind: the human-powered blimp.

The idea for such a strange vehicle emanated from Bill Watson, designer and constructor of unique models which perform like the real thing. Watson's ten-foot flying model of a prehistoric pterodactyl was seen in the film, *Wargames*. His interests range from mechanical robots to model blimps and many of these weird and wonderful devices have been seen lurking around Hollywood TV and movie lots for years.

Watson himself, is no *ab-initio* to the world of manpowered flight. He built the world's first human-powered aircraft called the Icarus HPA-1 which, although unsuccessful in achieving Kremer prize status, was certainly the first to fly by man's pedal power alone. Watson later joined forces with Dr. Paul MacCready as chief constructor on the cross-Channel Gossamer Albatross.

Since 1968 Watson has been fascinated by blimps. He built several radio controlled models and took first place in the International Modelers Show for two years running with his "Gold Football." A cable television entertainer named Gallagher heard of Watson's projects last year and asked him to build a blimp as a prop. The radio-controlled construction was to resemble a watermelon and, with its success, the comic commissioned a human-powered flying machine. Watson, of course, saw no problem despite the fact that his blimp experience was limited solely to models. As a pioneer in human-powered flight, few were better qualified to develop such a project.

Gallagher, whose trademark is the unkept long-haired hippie look of the late 1960s, races around frenetically wearing Beatlemania trousers and sloppy-Joe tee shirts. He latched onto Watson's concept in order to assist his routine and provided the necessary $40,000 to make it happen—a figure Gallagher claims he can earn back in a show or two.

The comic boasts a degree in chemical engineering but dropped out of the profession after appearing at his Allied Chemicals job in Chicago wearing a pin striped double-breasted suit with a tommy gun in hand. His superiors found little comedy or apparent originality in the routine. Gallagher commented, "I don't seem to take things seriously." Summarily fired, he looked for greener pastures where misfits could more easily thrive. Cable comedy looked a more palatable profession than chief chemist at a fertilizer factory. Interestingly, Gallagher's latest endeavors are not that far afield from his early pursuits; he is still earning a living pushing manure around—in the kindest sense of course. He estimates that he conducted 175 live comic performances last year with numerous appearances on Showtime cable specials.

Although the airship was designed as a prop for Gallagher's shows, the comedian does have additional plans in mind. He anticipates using the blimp in conjunction with his ongoing fight against air pollution, bringing public attention and awareness to some of the more serious violators. This effort may be purely fanciful self-aggrandizement considering the pure impracticality of his blimp effort.

White Dwarf hangs lazily in the sky as Gallagher demonstrates his airship prowess.

In a semi-jocular manner he talked of flight parks where one might rent a White Dwarf in one place and pedal downwind to another—not the kind of sport to be looked upon kindly in cities such as Los Angeles, Chicago or New York with their dense air traffic. Imagine the aerial disruption caused when a runaway blimp accidentally crosses the final approach course to LAX, JFK or ORD. Better still, the havoc created with metropolitan helicopter operations as they confront uncontrollable blimps and seek evasive action over populated areas. The one certainly about dreamers is that they aren't necessarily realistic.

There is little doubt that the blimp can well serve Gallagher's public image, but not by virtue of anticipated news media coverage for his purported altruistic public service efforts. As an Ultralight, the blimp is severely restricted and operationally limited to sparsely-populated areas where Gallagher's public service plight will go largely unnoticed. Use of the blimp for any real benefit to Gallagher will most likely be in conjunction with his show rather than on-the-spot news dealings with blatant atmospheric polluters.

The blimp, of course, could be licensed allowing penetration into significant dense areas of population where pollution violators are more likely to be media grist, but then pedal pushers would require full airship ratings. Currently only about twenty U.S. pilots, all associated with Goodyear Airship operations, are so qualified and most unlikely

MUSCLE POWERED BLIMPS

MUSCLE POWERED BLIMPS

WHITE DWARF SPECIFICATIONS

Powerplant	One humanoid
Length	48 feet
Height	27 feet
Width	15 feet
Envelope volume	6200 cu ft
Lifting gas	Helium
Seats	One
Empty weight	140 lbs
Gross weight	390 lbs
Pilot weight	70-245 lbs
Minimum speed	0 mph
Cruise speed	8 mph (0.3 pedal horsepower)
Maximum speed	14 mph (0.6 pedal horsepower)
Range	Limited by pilot endurance
Fuel capacity	0 gals
Envelope type	Raven Industries TIF 6000 (modified)
Propeller	64 inch diameter with ground adjustable pitch. Spruce/foam construction— Produces 12 lbs of cruise thrust

candidates for Gallagher's efforts.

Watson's only practical blimp experience developed in the construction of several radio controlled models, some of which grew as long as fifteen feet. He spent about 4½ weeks on design while consulting a number of engineer associates for stress analysis. Construction began in June 1984, requiring a short 2½ months for completion. Watson,

Test Pilot Bryan Allen supervises first time pedal pushers as they clamor around the patch.

Allen and Charlie Sink were involved in the total effort. The White Dwarf, as it is called by Gallagher, was obviously a major undertaking, some four times larger than anything else previously conceived but, according to Watson, it was merely a question of scale.

The man-powered blimp has proven much safer than expected and Allen feels it is a true every-man machine. He compared it to his other human-powered flight experience, "The Gossamer Condor and Gossamer Albatross were really good at winning prizes for human-powered flight," he said, "but they really weren't accessible to the ordinary person. They cost two or three or five times as much as this blimp and they required a certain amount of athletic ability to keep them in the air. As you can see the blimp requires no athletic ability whatsoever. It really is a whole new realm in terms of human-powered aircraft."

According to Watson, the design criteria were based on something simple to build, simple to repair and something with certain aesthetic style. "It didn't have to be shaped the way it is," he said, referring to the fuselage. "There are a couple of things more complicated than necessary." Allen added, "It's Bill's interpretation of what a semi-frivolous device should look like—this makes it a bit more fun. If you want to go hi-tech you'd use a single carbon fiber tube rather than the aluminum tubes and trusses."

From a construction standpoint, the semirigid blimp is fascinating. The bag itself is a stock item made by Raven Industries and constructed of five ounce per square yard nylon with a polyurethane coating. Raven produces this in a series of advertising blimps—tethered gas bags shaped much like the barrage balloons employed over London during World War Two. Ravens presently carry advertising slogans instead.

The White Dwarf is actually a slightly modified Raven TIF 6000 advertising balloon. Modification involved removal of the standard adverblimp tail fins and some of the load patches. Watson noted, "We chose not to have fins for weight saving but also stability requirements. We wanted less directional stability so that the bag could turn more easily." Fins are generally required for weather vaning into wind—an undesirable characteristic if one wants to go in another direction without being at the mercy of low level gusts. The blimp is sensitive to winds as little as two to three miles per hour.

The gondola, constructed from aircraft grade aluminum tubing, is designed with a load capacity of four to five Gs. Any one supporting line to the helium bag (there are sixteen in total) could carry the entire structure. Each load patch is stated to be good for a 400-pound strain. Total empty weight for the airship is 140 pounds including the gondola and bag. Since this Heath Robinson contraption is indeed powered by one human, it still falls well within the 254-pound weight limit for powered Ultralights.

By incorporating a large 90-square-foot rudder, reasonable turns are possible. Natural balloon inertia prevents any major yaw tendency once a course is selected. "The idea was to be able to fly in indoor auditoriums or for some of Gallagher's shows," mentioned Watson. "We wanted something as maneuverable as possible."

Rather than normal pitch control in the form of an elevator, the White Dwarf utilizes vectored thrust from its pusher propeller. Linked to a parking brake handle, three detent positions allow an upward sweep of 40 degrees from the vertical for descents and a downward sweep of 60 degrees from the vertical for ascents. The gondola platform remains relatively level during these maneuvers eliminating the steep pitch attitudes associated with most blimp operations. Of course human pedal power is limited to about ½ hp on average so the chances of obtaining any severe pitch changes are nil.

Cruise speeds of 8 mph can be

achieved with an output of 1/3 hp. Double that, a feat attributable to only the best trained cyclists, will provide a forward motion of 14 mph. At about 18 mph the balloon will begin to pitch, oscillate and go dynamically unstable. According to Allen, this would require a greater than 1 hp output—unlikely even for a bionic bicyclist. On his Channel flight with the Gossamer Albatross, Allen, who was extremely well-trained for the event, developed only an average 1/3 hp for the duration.

A winch type handle controls the rudder position and a rear view mirror confirms its actual setting as well as the position of the thrust vector. The rudder is hinged to expose a large balance tab area enabling a more effective turn using propeller thrust as well as normal slipstream effect. Since the craft will normally cruises at such low forward speeds, propeller thrust against the large airfoil is essential to induce a turn.

A combination pressure relief valve and gas filler valve is located near the pilot's left hand. Relief is automatically provided when internal pressures reach more than three inches of water. This will prevent the balloon from bursting when large altitudes or temperature changes take place.

Generally, helium pressures are measured in inches of water since mercury readings would be very small. In the 6200-cubic-foot White Dwarf, for example, normal inside pressures are about 1/3 to 1/2 inch of water. This corresponds to about a .02 pound per square inch pressure differential between inside and outside. Where one might expect the gas to fizzle out quickly such as in a child's balloon, the pressures are so small that a puncture would hardly be noticed operationally. A bullet hole would involve only a slow leak and certainly not impair a safe landing.

The blimp is usually flown at neutral buoyancy or possibly one pound heavy. This means that its all up weight is always somewhere in the area of 390 pounds. Total payload, which includes pilot and ballast must, therefore, also remain the same—about 250 pounds for neutral stability. If the pilot weighs 200 pounds, 50 pounds of additional weight will be required. Two small water ballast tanks, located fore and aft, are provided for this purpose and include toggles to allow jettisoning if necessary. A heavy condition, caused by flying into cooler temperatures, may require some water venting but, on the ground, water venting is used for fine tuning to an optimum lift-weight relationship.

In addition to the water tanks, four fuselage positions provide for placement of 16 lb lead weights. These are designed to fit into the triangular lattice structure and are then held by nylon straps and velcro fasteners.

Because altitude changes are minimal (less than 1000 ft), ballonets used in large production airships have been eliminated. The ballonet is an inner airbag which absorbs helium displacement, expansion and contraction as large altitude or temperature changes take place.

A single pilot seat is fore and aft adjustable so that leg length can be matched to pedal position. The large diameter main gear incorporates a metal chain which then connects to a 4-to-1 ratio step up gear which drives the propeller via a plastic chain. The propeller is mounted on an aluminum pylon with a swivel head to allow the three vector positions mentioned earlier.

The blimp can be controlled in only two axes: yaw, using rudder, and pitch using vectored thrust. Pedal power will allow a forward maximum speed of 10-12 miles per hour. Reverse thrust can slow, stop and cause backward motion. However, the propeller is only about 60-70 percent effective in this mode.

Because the blimp itself has highly positive inertia, maneuvering in confined spaces is difficult and a wind of any description adds to the complexity of the situation. It was fascinating to sense the effect of a 2 mph zephyr which would go almost totally unnoticed while standing on the ground.

Initially the White Dwarf carried a redundant power source in the form of a small steerable electric motor with 2½ lbs thrust. This was a safety measure in case the prop fell off or the drive chain broke. The team has removed this only because of the inherent safety which has been documented. The current tethered flights prevent any mishaps but later free flight tests may demand the safety motor replacement.

In practice, the blimp is kept inflated and hangared at Camarillo Airport, California. The $600 worth of helium required to fill the 6200-cubic-foot capacity is a deterrent to deflating and carting home. Hangared at zero inches of water (inside pressure equal to outside pressure), only 15 to 20 cubic feet of helium are lost per day. That amounts to a dollar or two in top-up cost.

So far airport operations have not been hindered in the slightest. The blimp is flown with a nylon tether line so that in any unforeseen circumstance it can be reeled in like a giant fish. Flights to date have remained below the normal airport traffic pattern and have not reached altitudes above 150 feet or extended beyond a 150 yard radius. Even in such limited operational circumstances the blimp is a satisfying delight to fly. Plans are currently in development for an extended free flight in California's barren San Joaquin Valley.

Watson foresees production possibilities for his machine but he may have his head in the clouds when one considers the cost/utility equation. "If we do start manufacturing," he said, "it will be on a limited production basis as an experimental aircraft in kit form."

Although the theoretical and aerodynamic developments must be regarded in the highest order, the commercial justification is certainly not airtight. Initial costs were in the $40,000 bracket which includes design and labor but Watson sees the blimp definitely cheaper once in production—probably in the area of $15,000-$20,000. Unfortunately the Raven-made bag costs nearly $13,000 plus another $600 for a helium top-up. On the other hand, maintenance is only a couple of dollars a day so, *if* you had a place to put it, operating cost would be less than a typical Ultralight. The price is steep, but this certainly won't deter Arab Shieks and Jacqueline Onasis types —of which there are plenty about. Imagine "blimping" silently past the Pyramids at dusk . . . or over your Loire Valley chateau? Malcolm Forbes where are you?

Watson and Allen are considering this a serious pursuit. Certainly their contract with Gallagher may involve further developments even if production runs do not become a reality. Both are on Gallagher's full-time payroll developing the blimp concept as well as several other ideas for mechanical props. A ¼ scale model has been built using the smallest in the Raven adver-blimp series, a 110-cubic-foot bag. This is radio controlled since it is too small to hold a human payload.

Both Watson and Allen see the White Dwarf as a stepping stone to even more exciting developments: a two-seater which could be used for airship license trainees, a solar-powered one and even an around the world blimp. They feel that the comedian was the inspiration and

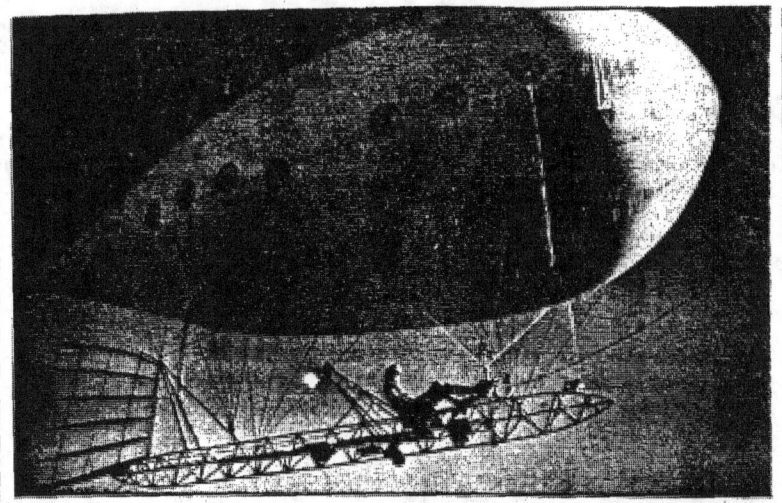

Cruising past the sun—Allen sees this as a possible around-the-world-machine.

Champagne send-off—Gallagher christens himself as Bryan Allen and Bill Watson prepare the comedian for the wild blue yonder.

incentive to make this happen and are considering a variety of future possibilities. Allen admits, "Most of the ideas such as flight parks are whimsical. Very few are practical other than for recreational purposes."

Is the blimp an alternative to conventional Ultralight flying? Allen points to the severe weather limitations. Although winds of 15-20 knots are fine for a fixed-wing Ultralight, the blimp has a maximum headwind component of 10 mph—about the top speed at which one can pedal. In such a case one could hover at best but be severely restricted in upwind movement.

One certainty is that blimp safety in calm conditions is far greater than most powered Ultralights. The first timer with no training can operate in complete confidence as long as he has nothing to run into, the tether line assists in this regard. Allen commented, "The overall safety aspects are mind boggling. Just put a person in and say, 'Go fly it.'" After seven days of "blimping" Allen had logged about fifteen hours. "With a fixed-wing Ultralight," he noted, "by the seventh day, you're muscle making ten-minute hops."

For theorists, developers and achievers like Allen and Watson, comedy is a long way from man-powered flight development. Does the record-breaking, human engine foresee learning this new trade? Allen was sitting in the driver's seat of the blimp just hovering some ten feet in the air, feet moving slowly to keep the White Dwarf in position. "I'm really not cut out for that," he said. In his thoughts are something far more significant. "I'm convinced that the best way to fly around the world, nonstop and unrefueled is in one of these—a solar powered airship with a crew of two or three. With solar power and battery storage you could probably fly around the world in 18-20 days."

As he back pedaled and turned the craft around, Allen smiled knowingly. Clearly, he was enjoying himself. Then, off he went. The loudest sound one could hear above that of some chirping and astonished birds was the honk of the horn and the ringing of the bell as he flashed into the distance pedaling at a heady 10 miles per hour. ☐

MUSCLE POWERED BLIMPS

Who is Bryan Allen?

The world's most famous pilot and engine of human-powered airplanes has returned to earth...sort of.

By Stephan Wilkinson

Hollywood knows more about formulas than Dow Chemical ever will, and it has poured forth a torrent of ritualized "youth films" with reassuringly familiar plots and characters—the Jock, the Nerd, the Valley Girl, the Misunderstood Greaser. There are parts in all of them for the young Bryan Allen, who could play the lead in *Breaking Away to the Future II,* a film set in a small farming community in central California. Allen would be the skinny, slope-shouldered, near-sighted kid with the schoolteacher father—an indignity any preacher's or teacher's child will confirm—who spends his time pedaling a 10-speed while the dudes are tuning their GTOs. ("I won the Homecoming Day bicycle race three times in a row," he today muses. "It was the one day a year I was normal: *everybody* was riding a bike.")

Allen then goes off to the California Institute of Technology—Nerd State, through celluloid eyes—to become a biologist, and he takes a job testing fertilizer on a huge corporate farm in the San Joaquin Valley. But the last reel hasn't unwound: he gets sidetracked and goes to work for a weird inventor named Paul MacCready, who comes up with history's first successful human-powered airplanes! And Bryan Allen, pedaling with the power of a linebacker, flies them

His pedal-plane fame may be fleeting, but Allen's joy in creating things, even bubbles, hasn't burst.

By pedaling the Gossamer Albatross *across the English Channel in 1979, Bryan Allen flew into history.*

into the pages of history.

He is featured in *Time* and *National Geographic,* gets a brass plaque from his high school honoring him "For His Contributions to Aviation" (in your face, Mister Touchdown), ends up on the lecture circuit, and gets the girl. It's a wrap. And it's all true. Allen did indeed grow up in a social sidestream in Tulare, California, a town that could have been the set for *American Graffiti.* And he did become pilot, powerplant, and one of the builders of the record-setting *Gossamer Condor* and *Gossamer Albatross.*

On August 23, 1977, Allen permanently widened his and aviation's horizons. Pedaling the shimmering *Gossamer Condor,* he completed the first human-powered flight over the mile-long course stipulated by Henry Kremer. For 18 years the British industrialist had offered an award—finally totaling $86,000—for exactly that simple-sounding but in fact incredible accomplishment. Birds do it, bees do it, but not until MacCready's innovative engineering was matched to a superb engine—a 140-pound bicycle racer and hang-glider pilot who could produce a consistent one-half horsepower for minutes on end—did a human being do it.

"MacCready said, 'Let's build an airplane to win the Kremer Award,'" Allen explains, "rather than, 'Let's build an absolutely wonderful human-powered airplane.' While the British teams were trying to bludgeon their way to success through sophistication, MacCready chose the cheapest and easiest course. He was willing to build an extreme machine that would be very limited and impractical but that *would* win the prize." In a sense it was exactly the route that Charles Lindbergh had chosen when he created an unstable, im-

MUSCLE POWERED BLIMPS

Who is Brian Allen?
cont.

practical, single-seat, nearly windowless airplane that could be constructed quickly with but one mission in mind.

Two years later, in June 1979, Allen spent an agonizing two hours and 49 minutes pedaling MacCready's next beach. He won the MacCready team a second Kremer award—this one worth $210,000 for the first human-powered flight across the English Channel—and entered the pantheon of competitors who have asked more of their muscles than seems humanly possible.

Allen, who may be the most unassuming super athlete in the history of sweat, has had no problem turning his back on instant notoriety. "It was fun dealing with fame for about a year, which was all I intended before going back to obscurity," he says. "I felt I'd taken time off—not accomplished anything, not built anything."

After a brief career as the public-affairs manager of a company making ultralight airplanes in New Mexico, Allen went back to building things, at the Watson Model Works, a laid-back work-shop on a side street near Van Nuys Airport, in Los Angeles. There, Allen and Bill Watson, a friend from the MacCready days, make stage props for the comedian Gallagher, a prince of craziness whose popular act consists largely of word-play toys, food fights, and the launching of ludicrous missiles into the audience.

The Styrofoam flying objects—watermelons, fish, elephant tongues, a dollar-sign-shaped bird called an s-crow—flow seemingly unbounded from Allen's childlike enthusiasm. (in his little office, there is a congratulations-on-your-new-job card from a friend. It shows a young man riding an oversize tricycle, and the legend reads, "The secret of staying young is finding an age you really like and staying with it.).

There is something strangely fitting in Allen working at the beck of a comedian. Not only does he have a sense of humor that allows him to see absurdity in what others take seriously, but his own *Gossamer Albatross* flight danced around the edge of hilarity itself. Had Gallagher been the sponsor, we'd have fallen down laughing at the spectacle of a half crazed man in horn-rims and plastic helmet riding a Saran-wrapped bicycle across the English Channel.

From the ceiling and walls of the shop hang a dizzying variety of vehicles designed by Watson & Allen — from a pedal boat faster than most sailboats to a bicycle designed for up-hill races only—that look like they're waiting for the Three Stooges to come along and take them for a spin. But visible here and there amid all the goofiness are some of the components of an aircraft that weds Allen firmly if irreverently to his famous past. The White Dwarf is a 47.5 foot-long human-powered blimp, in which Allen has already set two world records: for flight distance and duration for small airships, *including* those with engines.

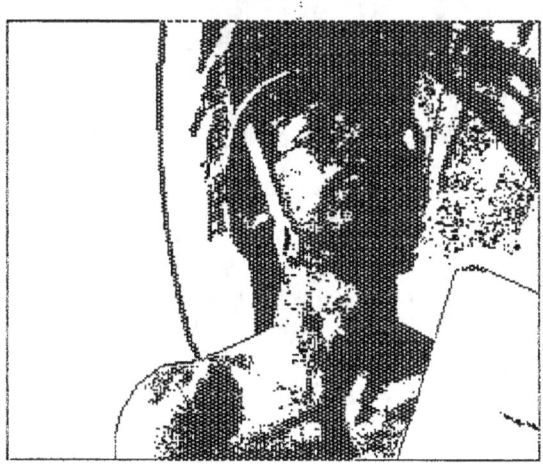

A superb engine for human-powered aircraft, the 140-pound Allen churns out a steddy one-half horsepower.

Unable to stay snugly earthbound, Allen helped devise the White Dwarf, perhaps the Model T of "sport blimps."

On their own version of a bicycle built for two, Allen ponders while his shop partner, Bill Watson, pedals. The two are friends from Gossamer *days.*

He powered the turnip-shaped craft from Thermal to Brawley, California, covering 58.08 miles in eight hours and 50 minutes. Allen spent the time pedaling lazily—"I was way out of shape," he says—and holding shouted conversations with baffled long-distance truckers on the highway that paralleled his route.

The White Dwarf has also helped convince him that his earlier human-powered deeds were perhaps misdirected. Henry Kremer's money was the carrot that led experimenters such as Paul MacCready to a remarkable variety of carbohydrate-fueled accomplishments, but the rules forbade the use of lighter-than-air gases for lift. "Isn't that ridiculous?" Allen asks. "It's as though somebody said, 'You have to travel from here to there, but you have to crawl.' Or as though somebody had a literary contest that would only accept hand-written entries on homemade paper."

The designers of the nearly 60 human-powered airplanes built since the Kremer awards were first announced might disagree, and the state of muscle-plane aviation advances because of them. MacCready has gone on to build superlight airplanes powered by solar energy. A team of students at the Massachusetts Institute of Technology built a craft that recently won a Kremer prize for speed, flying a 1,500-meter course at approximately 20 mph, using the pilot's leg-power as well as his energy stored in a battery via a pedal-driven generator shortly before the flight. The team is now organizing an attempt to fly a human-powered airplane from the island of Crete to the Greek mainland—the nearly 70 miles that were the downfall of Icarus and his feathered wings. And a West German team has built an airplane so sophisticated that it was able to best the M.I.T. team's speed record on pedal power alone, with no stored energy.

Allen sees the pedal-driven White Dwarf as the prototype of a new class of "sport blimps" that could make human-powered flight... well not exactly *practical*, at a potential $40,000 per copy and a bulk that requires renting a small hangar, but more attainable than it is through the use of fragile, labor-intensive aircraft such as the *Gossamer*s. The blimp is a robust little sausage, and its suspension cables, aluminum-tube keel, and fittings have all been tested to five Gs—about three times the strength of the flimsy craft in which Allen made history. Indeed, the White Dwarf has been flown by pedalers as heavy as 250 pounds and in breezes that would turn a human-powered airplane into a plastic tumbleweed. (He and Watson are also

developing a slightly larger two-seat sport blimp powered by a pair of fist-sized model-airplane engines.)

Growing up in Tulare, Allen would ride his bike to the US 99 overpass and watch cars flowing north past a sign that read "Vancouver, 1,017 miles." He would imagine that perhaps one of those very cars was actually making that journey. The White Dwarf is yet another reflection of the childlike desire for simple freedom and uncomplicated mobility that seems to have driven Bryan Allen ever since. ✈

BUILDING AND FLYING A HUMAN-POWERED AIRSHIP... WHITE DWARF

By Bryan L. Allen
12164 Emelita
North Hollywood, CA 91607

MUSCLE POWERED BLIMPS

Sometimes the best projects and greatest undertakings start just by someone asking "Why not?" Gallagher, a highly acclaimed comedian, saw a model blimp in 1983 that was built by a diversely-talented Southern Californian named Bill Watson, and decided on the spot he **needed** one for his stage act. But Gallagher didn't like the utilitarian style of Watson's demonstrator; he wanted one that would make people laugh. Scarce weeks after their first meeting, there's Gallagher on national TV guiding a Watson-built blimp whose bag looked like a watermelon (a Gallagher trademark), with a rudder that resembled a child's kite turned sideways, and a little model Gallagher, seated on an old-style bicycle hooked to a propeller, madly pedaling away. "Whattaya think," he asked his audience, "would you rather have one of these, or your car?" After mastering the art of radio-controlled flight, there were times when Gallagher looked at his blimp flying out over the heads of amazed crowds and thought, "You know . . . that looks like fun." So he went back to Bill with the question, "Can you build a blimp that will carry **me**?" Bill did some studies and talked with other experts to verify his conclusions about possible structures, regulations, materials, configurations. Little more than four months later, there's Gallagher pedaling around the sky with his new high-tech toy, the **White Dwarf**.

The **Dwarf** was designed by Bill Watson in about a month. Over fifty different configurations were considered and rejected. Good designing is always more complex than the end product indicates; for instance, one of the design criteria was that this pedal-powered airship pose the least possible hazard to spectators. Thus the propeller location, materials, and design had to minimize the likelihood of anyone being bonked on the noggin, and to reduce the consequences if such a thing happened anyway. A larger, sharper-edged prop like the 13-1/2 Kevlar and carbon-fiber wonder on the **Gossamer Albatross** would have been more efficient than the 5-1/2 foam-and-spruce prop used on the blimp, but safety dictated this reduction in performance. Bill discovered that Raven Industries in South Dakota made an envelope for suspending electronic payloads which could be modified to work; again, not an optimum solution from a pure performance standpoint, but one which allowed the blimp to be built in less time.

The **White Dwarf** took three people working 40-hour weeks 2-1/2 months to construct. As with any prototype aircraft, there were many details which had to be worked out during construction. The control system, the seat, the fittings . . . all these and more were invented from scratch. We even built a trailer to haul the deflated airship around, helped by Bill's brother Skip. Many samples of proposed components were built and ripped apart, under the documentative eye of a video-camera, with a hydraulic test-rig we too built ourselves. By following the clear and concise dictum, "Keep it simple, keep it light, with safety **first**", the solutions came easily. One thing that made the construction crew of Bill Watson, Charlie Sink and myself different from many homebuilders is that we prefer any project to take less time rather than more. "Good enough" was a phrase heard many times during the construction of the **White Dwarf**. Yet when an aircraft is pared down to essentials, beauty can still be present. Elegance in the placement of a rivet, or curve of a tube, can delight the eye of those who appreciate structures. We think the **White Dwarf** turned out pretty darned nice, thank you, and others have told us so, too. At a total cost under $40,000, which includes the trailer, tooling, and generous wages paid to the three of us, this is one of the lowest-cost "right the first time" ultralight prototypes of original design that I know of.

View from a nose-mounted camera from the *White Dwarf* pedal-powered airship, conceived and owned by the popular comedian Gallagher. The large rear-mounted rudder is controlled by the ball-topped lever seen in the pilot's right hand. Suspension lines are dacron-sheathed Kevlar®; any *one* of these seemingly frail lines (there are 24 all told) can carry up to 400 pounds without failure, more than strong enough to support the entire 315-pound maximum payload of pilot and fuselage. These lines fasten to high-strength carabiners (black in photo) which then loop through red-and-yellow mountain-climbing ropes, the ropes, tying to steel shackles bolted to reinforced points on the 2024-T3 aluminum-tube fuselage. Titanium cycling pedals drive a short segment of steel chain, which then drives a steel-cable and plastic chain at a four to one ratio; the plastic chain drives the propeller through a right-angle gearbox located at the top of the prop tower. The brownish-colored tank below the pilot is one of two fiberglass water-ballast tanks carried. The black circular object in front of the pilot is a rear-view mirror, used for monitoring rudder position; it is next to the airspeed meter. The blimp's shadow can just be seen along the right-hand side of the picture. Pilot: *White Dwarf* designer Bill Watson. Location: Camarillo, Calif. Photo © 1984 Bill Watson/Wizard of Odd, Inc.

CALOI PEDAL-POWERED AIRSHIP cont.

Dados Tecnicos:
Dimensoes do Dirigivel:
. Volume do Bojo: 275 m3
. Comprimento: = 1 7,5m
. Diametro: 6m
. Altura: 9m
Dimensoes da Gondola:
. Comprimento: 8,66m
. Largura: 0,47m
Sistema Populsor:
Propulsao humana com transmissao a pedal e helice orientavel, atingindo a velocidade de ate 15 km/h
Material Utilizado:
. Bojo: 300 m2 de poliester revestide de poliuretano
. Gondola: aluminio de alta resistencia
. Sistema de transmissao e movimgntacao da Helice: por corrente.
. Helice: madeira de balsa entelada

Technical Data:
Dirigible Dimensions: (9711 cf)
  Envelope Volume: 275 cu.m.
  Length: 17.5 meters (57.41')
  Diameter: 6 meters (19.68')
  Height: 9 meters (29.42')
Gondola Dimensions:
  Length: 8.66 meters (28.41')
  Width: 0.47 meters (1.541')
Propulsion System:
  Human power driven propeller
  Vectored transmission
  Speed target: 15 kph. (9.32mph)
Materials used:
  Envelope: 300 sq.m. Polyester polyurethane coated
  Gondola: High strength Aluminum
  Drive System: Bicycle chain drive into a right angle prop gearbox
  Prop: Balsa wood

Equipe envolvida no projeto e construcao: **Project Constructors**:
Grupo Neptun. Karl Gruschka, Stephan Santos Klasta; Nelson Barbosa Correa; Marco Antonio Martin Migliano; Luiz Antonio V. Cerciello*, Enos Tangerino; Colabordors: Gino Dominicini, and Rafael Mora-Neves*.

Editors NOTE: As a Brazilian National, and a resident of Sao Paulo, I am very familiar with the *Neptune Group* from its inception, and am still in contact with some of the original organizers (*). The group was basically disbanded after the last flight demonstration in 1989. Unfortunately, a key element in planning was omitted (where do you put it after inflation?), and while the problem was being discussed late in the day, the wind solved the delima. The $ loss of the helium was the end of their financial resources, and each member went on their separate ways, to bigger & better things in life (like eating regular).

The mechanical portion(s) accompanied Karl G. back to Germany, and at this writing, has not reappeared in any media. Karl is a very good engineer, and no doubt, will resurface in the LTA community eventually.

Robert J. Recks, engineer/aeronaut
1 June 1998

The editor is solely responsible for the translation of this article. It is hereby submitted to the Directors of *AVIACAO em REVISTA* for permission to republish this important information. It is hoped that the new generation of sport airships will take hold in the near future.

ROBERT J. RECKS; 1148 Third Avenue #44; Chula Vista, CA 91911-3149 USA
zacairships@compuserve.com

MUSCLE POWERED BLIMPS

Since Gallagher is a non-pilot, we intended to fly the blimp under Part 103 guidelines. By calling it an ultralight, we came up against the hostility many aviation folks have against machines having "no license required". We needed a **big** hangar to put the blimp in if we were to be able to leave it inflated for any period of time, but found that there are not too many airfields around with big hangars that welcome ultralight pedal-blimps, at least in the LA area. A hangar was a necessity for our first tests, since helium is not cheap and the **White Dwarf** would take over thirty welding-tankfuls of it at a cost of $800 to $1200. That's a lot of money to throw away just because of something unexpected, like bad weather. The solutions to our searches coalesced at a former Air Force base now operated by the County of Ventura called Camarillo Airport. A big-enough hangar, with folks glad to rent us space, was here, with "Air Camarillo — Hangar One" emblazoned on the side. The airport manager, Jim O'Neil, at first must have thought we were a bunch of looneys, but ended up proposing we use a site under the base leg of the pattern scarce yards from our hangar that was **perfect** for the test-flights on tether we wanted to perform. No wonder there were so many experimentals scattered around the hangars and ramps; why, there was even a field for fixed-wing ultralights down the way. We decided this place was **friendly**; it even had a good cafe!

We have found that operating a small airship (just over 6000 cubic feet volume, 27 feet tall, and 47 feet long) is easier than we could have ever hoped for. Turns inside a seventy-five foot circle can be accomplished, something we showed decisively by doing 180 degree turns inside the hangar a couple of days before our first outdoor flight. True "hangar-flying" is possible with such a small airship; something as big as the Astrodome, why, you could put several pedal-powered airships inside there and have races!

Response from the vectored-thrust system is better than anticipated; once outdoor flights commenced, we found that moderate pedaling would net 200 fpm climb rates and 300 fpm descents. Contrary to our expectations and experience with other human-powered aircraft, if a pilot is strapped in the seat it takes only **one** person to ground-handle the airship in winds as strong as ten to fourteen mph. And talk about easy to fly! After we put a mirror on the front pedal-mount post so pilots could see the rudder position, not one of the over forty people who have since flown her has had any trouble adjusting to the control system. "How hard do you have to pedal?" is answered with another question: "How fast do you want to go?". Pedal like a sixty-year-old office worker with a heart condition and you'll coast around at four or five mph; pedal like a bike racer going for a gold medal in the Olympics and you'll zip around (relatively speaking) at eleven or twelve mph. Gallagher saw a spectator one day looking particularly wistful while watching the blimp fly. "Come over here and get in this thing!", he called. The spectator, startled to have his dreams answered so directly, got in the seat looking a bit overwhelmed. I asked him if he had any flying experience. "Not a bit," he replied in a hushed voice. "Great," I told him, "you won't have any trouble. This lever makes you go up and down, this makes you turn, see the rudder in the mirror there? If you stop pedaling, you just float. Ignore these guages here, you won't need them. Have a good flight. Up you go!" He flew around for fifteen or twenty minutes with an ever-widening grin on his face; we finally had to haul him down so others could fly. I've soloed quite a few people in fixed-wing ultralights, but never completely gotten over the stomach-churning tension of watching a student make his first solo takeoff. I've **never** been tense soloing persons in the **White Dwarf**, even if they haven't ever been up in a flying machine before and have only been given thirty seconds of instruction. When we let any person fly the blimp for the first time, we dangle a "guide-rope" over the side which has five two-pound weights attached along its length. As the guide rope contacts the ground, the weights no longer pull down on the blimp so it then has an excess of lift and rebounds back away from the ground. No hard handlings possible when you have a landing gear with twenty-five feet of travel! Training radios? Don't need 'em; just shout up at the student. Student going the wrong way? Give a tug on the light but strong Kevlar tether rope and they come back to you whether they want to or not. An ultralight airship is an anxious ultralight flight-instructor's dream come true.

This smallest of existing airships poses a whole bunch of "Why nots?" to those in aviation. For instance, the most efficient human-powered airplane ever built, the **Gossamer Albatross**, proved in flight trials to be beyond the capabilities of most people for anything other than sub-one-minute flights bare inches above the ground. Though it made some amazing flights, the **Albatross** was good for little else than winning prizes. On the other hand, the **White Dwarf** has flown many flights over thirty minutes with non-athletes doing the flying. It is the first human-powered aircraft with the capability of being safely flown several hundred feet above takeoff, having an ultimate load-rating of about 5 Gs. Many spectators asked those involved with the Mac-Cready Gossamer-series aircraft why the wings weren't filled with helium. The answer was simple, but gave little satisfaction: the rules for the flight-prizes established by the English didn't allow it. It turns out that these Kremer prize-rules, which brought about the development of many amazing airplanes, many have been slightly misguided if their intent was to further interest in human-powered flight. For muscle-powered flight, I feel blimps are: safer, easier to fly, cheaper to build and operate, lower in maintenance, can be flown at more locations and tolerate weather and turbulence much better than fixed-wing craft. In other words, blimps are just hands-down **better** for pedal-powered flight.

What's next for this pedal-blimp? In mid-February of this year, we took the **White Dwarf** to the Palm Springs area for its first off-tether flights. We learned an enormous amount, about which I hope to talk in a future article. A summary of accomplishments: climbs to over 1000 feet AGL; a flight duration on one day of over eight and three/quarters hours; a cross-country distance travelled on the same day of nearly 100 kilometers; many hours of fun, floating scarce feet above date orchards and desert sage. But what else can we do with the concept? Air show appearances, possibly. Create a sister ship? I hope so (if for no other reason than this country **needs** to see a good blimp-pull now and then). Fly to work? Get serious. Parking lots around Los Angeles are crazy already. But no matter what we ultimately end up doing with the **White Dwarf**, I expect that once people realize how easy and how much fun a small blimp is to build and fly, there will be more pilots joining us in the joys that only small airships can provide.

ABOUT THE AUTHOR

Bryan L. Allen (EAA 147881, UL 1536), born October 1952, first learned to fly in 1974 on a foot-launched hang glider. He was the pilot of the **Gossamer Condor** pedal-powered airplane that won the original Kremer Prize in 1977 and which is now on permanent display in the National Air and Space Museum. He piloted the **Gossamer Albatross** across the English Channel in 1979 to win a prize of $200,000 for the group headed by Paul MacCready. He flew the backup **Albatross** at the EAA Convention in 1980. He holds the current FAI records for distance and duration in Class 1 (man-powered aircraft) and is the current record-holder in the Kremer 100,000 Pound Speed Challenge (flying the Bionic Bat around the 1500 meter course while averaging 23.45 mph). He holds private pilot ratings for both Airplane Single Engine Land and Gliders. Formerly with American Aerolights, he now works full-time building stage and screen props (**White Dwarf** among them) for the popular comedian Gallagher.

WHITE DWARF
specifications

Year built: 1985
Designed and built by: Bill Watson
Built for: Gallagher, the comedian
Dimensions
    Envelope length: 48'
    Maximum diameter: 17'
    Volume: 6200 cubic feet
    Overall Height: 27"

Weights:
    360 Empty: 140 lbs.
    Pilot weight range: 90-250 lbs.
    Maximum take-off weight: 390 lbs.

Performance:
    Maximum level speed: 12 mph
    Average cruising speed: 6-7 mph

Construction:
    Envelope: Helium filled, non-rigid polyurethane coated nylon.
    Helium maintained at average slight pressure of 0.02 lbs/sq, in. No ballonets.
    Fuselage: Open framework structure of 2024-T3 and 7075-T6 aluminum,
        stressed to 4+ G's, attached to the envelope by
        24 Dacron sheathed Kevlar lines.
    Ballasting: two water ballast tanks with pilot-controlled drains and up to 60 lbs.
        of lead ballast to control buoyancy.
    Venting: A three-inch diameter plug alongside the seat can be lifted by the pilot to
        vent helium. It opens automatically at a pressure of 4 inches of water.
    Aerodynamic controls: Large Mylar covered styrofoam and spruce rudder.
    Passengers: pilot only
    Power system: Pedal power, producing approx. 10 lbs. of cruise thrust
        via 4:1 gearing and plastic chain drive to a two-blade pusher propeller
    made of spruce and Styrofoam.
    Propeller can be vectored up and down for vertical control.

Misc.
    Cost to fill: approx, $1000,00
    FAI World for Duration, Class B Airships, BA-1 through BA-10,
        achieved by pilot Bryan Allen in 1985.

For more detailed information and photo archives, log onto:
        << http://www.teteport.com/-reedg/whitedwarf.html >>
Or contact the year 2000 owner: Reed Gleeson

MUSCLE POWERED BLIMPS

A reprint from *MODEL AVIATION* May 1985 p80
(White Dwarf)

For historical reference only

PHOTOS
PICTURE CREDITS: Bill and Phyllis Warner, except as noted.

In both the big picture (1) and the inset (2), the designer of "The Little White Cloud," Bill Watson, is at the pedals and controls. In the big picture, it's in front of (not over) the Constellation. (And remembering the recent article, "Beautiful and Ugly Airplanes," is there a more beautiful prop driven transport than the Constellation?) Watson has the rudder hard over in the inset (3). Rudder is controlled via a cord and a crank (4).

The van delivered a cylinder of helium to top off the airship (5). One of the hazards of buying a pig in a poke: the tank proved to be empty! Water ballast is added to the front tank so that the ship will weigh about a pound (with each individual pilot) before

Bill Watson gets ready for a flight (6). Design of his human powered blimp followed many years of experimenting with rubber- and electric-powered models. He also builds racing cycles.

Bill Watson rides as the airship is walked to its flying site at the Camarillo Airport (7). He's riding because the ship needs ballast. A novice can fly it after a brief instruction period.

Bill Watson (L) adjusts ballast for Gallagher, about to take flight (8). Gallagher is a successful TV comedian and philosopher who wants people to take notice of his "Little white Cloud" and stop polluting the air. Brian Allen (of Grossamer Condor and Albatross fame) calls the ship the "ultimate trainer".

Bill Watson turning the blimp (9). It is kept tethered even though flying free, "just in case." The flying site was between hangars and parked aircraft, so even the tiniest breeze could present a significant hazard.

Watson flies overhead (10). Note pyramidal water ballast tanks and helium filler tube which runs down to the pilot in case of need to release gas. Bill and Phyllis Warner pics except as noted.

The better half of the Warner team, Phyllis, flies the blimp overhead (11). Easily visible are the three transmission chains and one of the fiberglass pyramids containing the water ballast. Her right hand is on the rudder crank. A rear view mirror shows the rudder position. The bag hanging on the side holds the equillbrator (rope with weights attached along its length) which can be deployed to ease landings by novices- Total fuselage weight is about 60 lb.

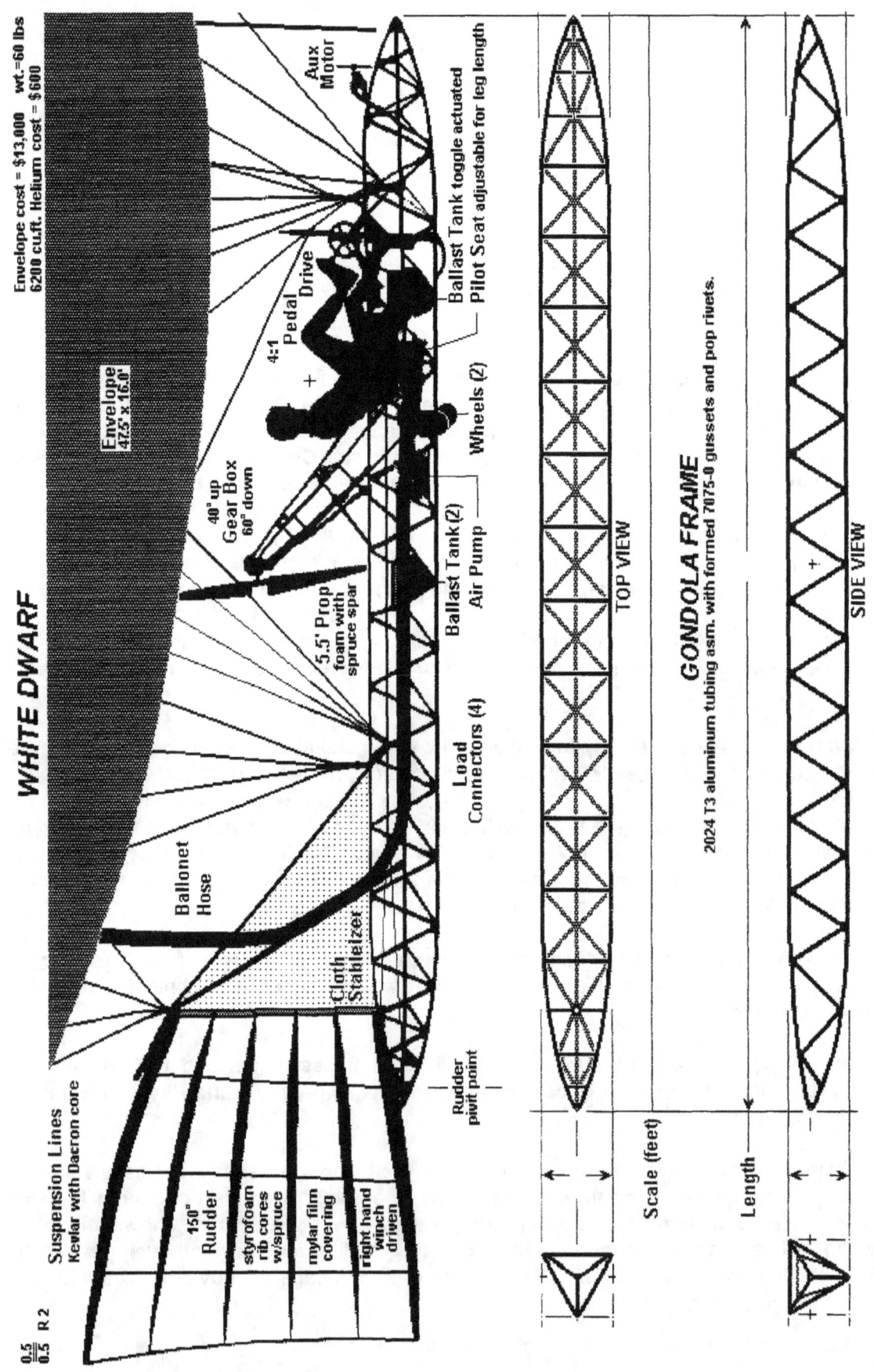

MUSCLE POWERED BLIMPS

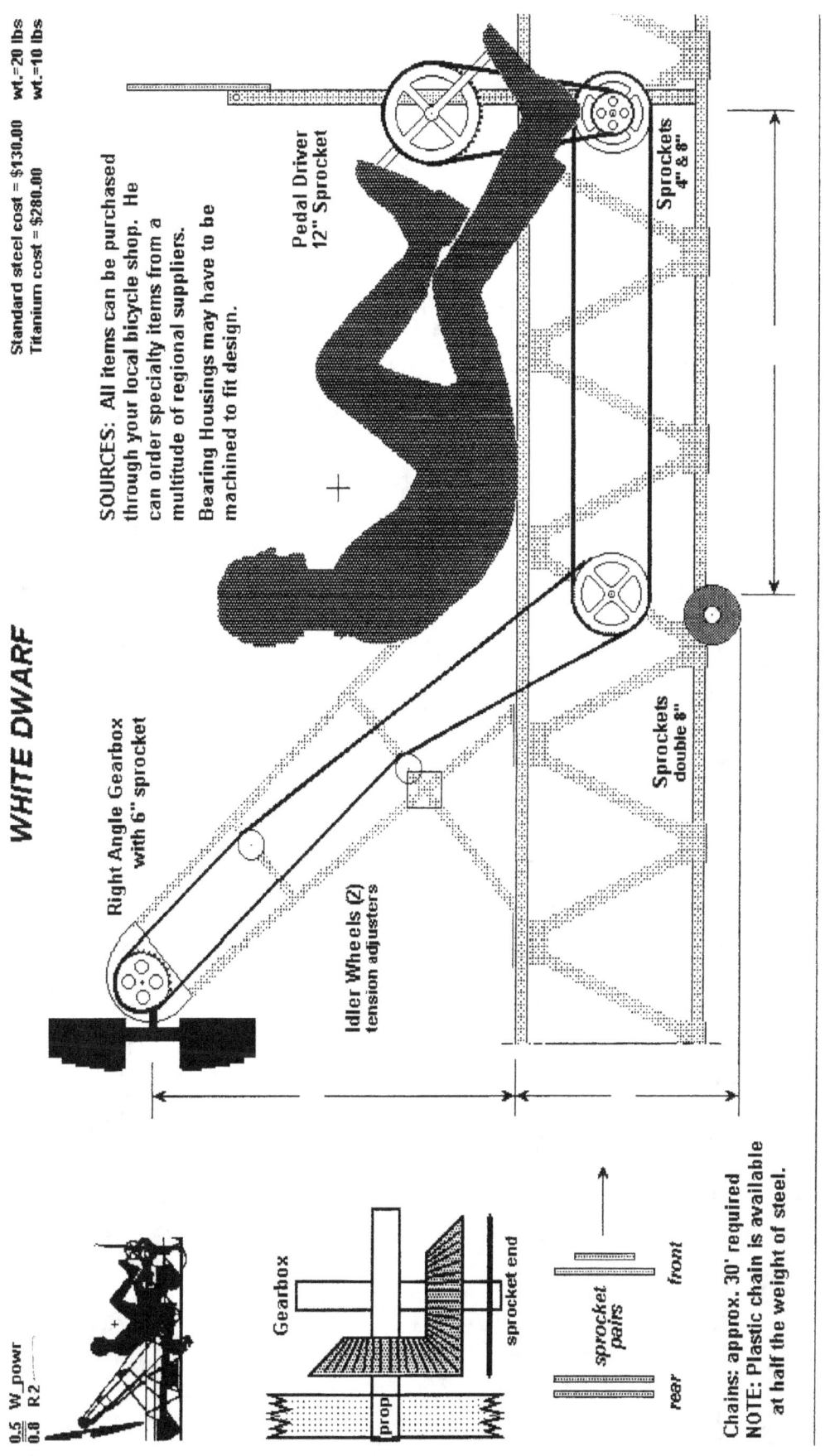

MUSCLE POWERED BLIMPS

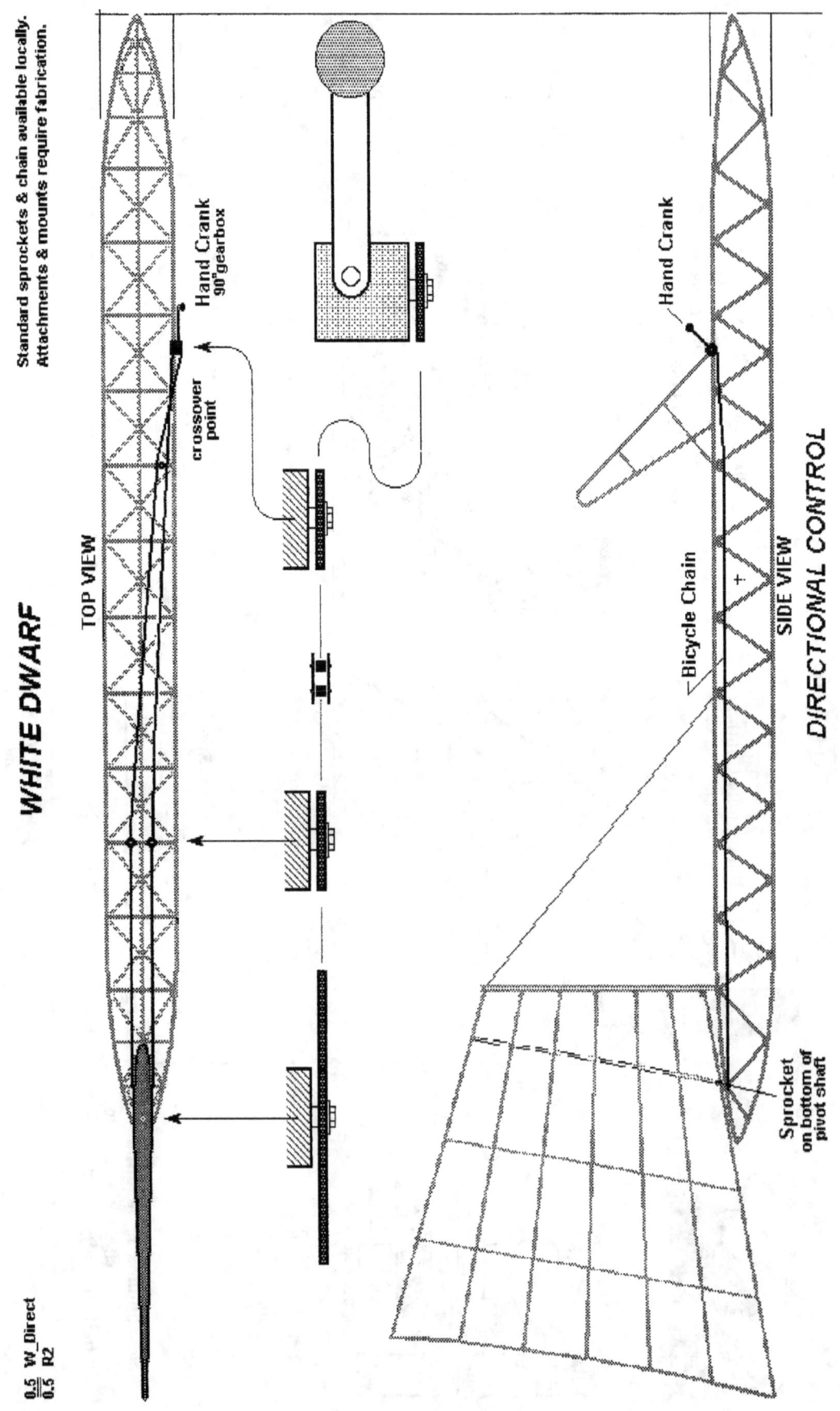

MUSCLE POWERED BLIMPS

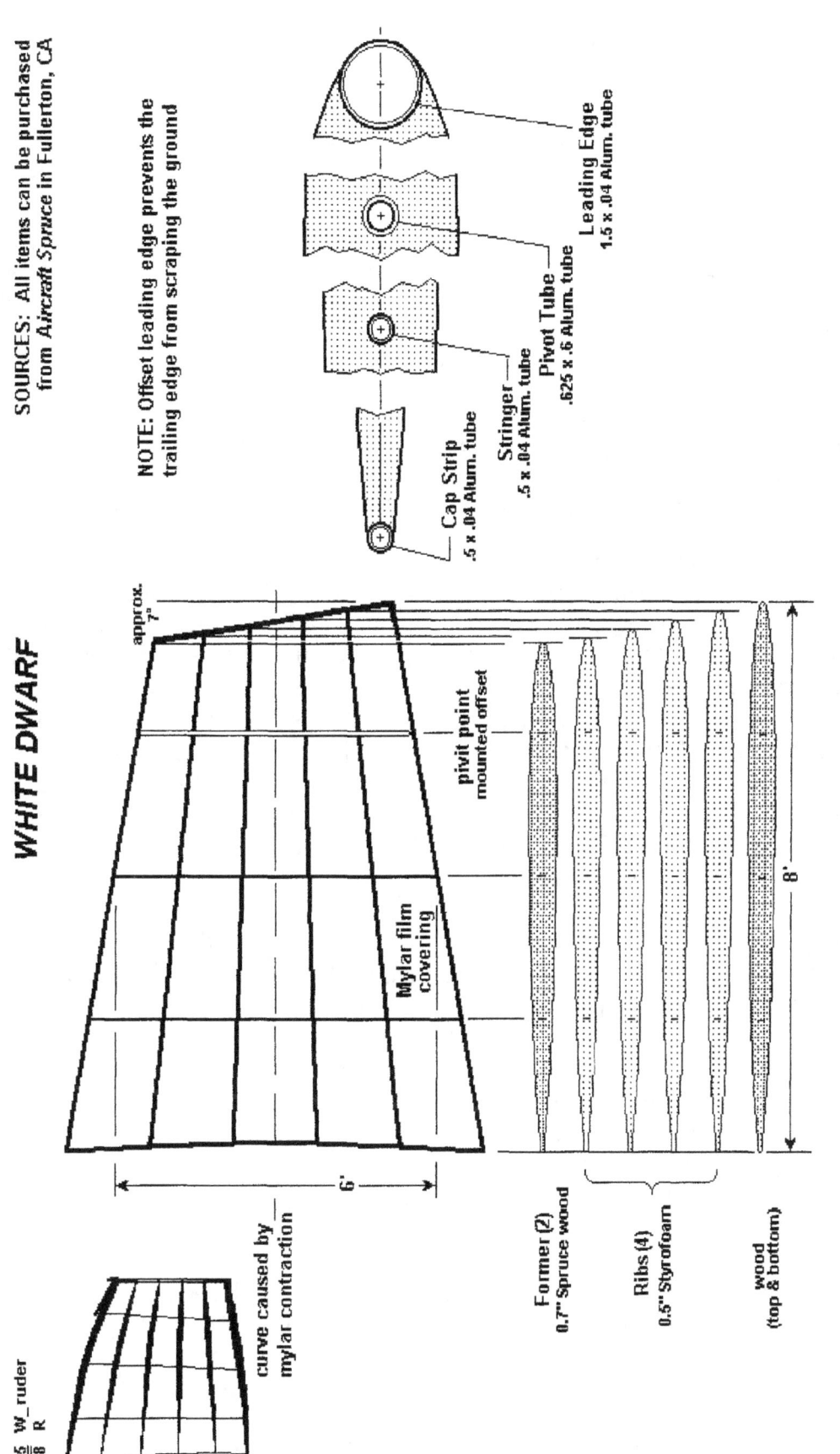

WHITE DWARF

RUDDER CONSTRUCTION
aerodynamic airfoil with offset pivot

SOURCES: All items can be purchased from *Aircraft Spruce* in Fullerton, CA

NOTE: Offset leading edge prevents the trailing edge from scraping the ground

177

What's up with the blimp? (24 Jan. 2001)

The White Dwarf Flies Again!

It's the White Dwarf (WD), built by Bill Watson and others for the comedian Gallagher in about 1984. This beautiful and innovative pedal-powered airship was used by Gallagher to do a Showtime special, Bryan Allen flew it 58 miles to set an FAI world record for that class airship, Bill and friends took it to Oshkosh, and then they put it behind Bill Watson's parents' house, where it stayed for about 14 years.

Reed Gleason (present owner) had been enjoying various ways to fly, and noticed that the slower he went, the more fun it was. He decided he wanted a blimp, and contacted Bill Arras, an acquaintance from serious hang gliding days. Bill Arras wanted a blimp too, and set about finding out how to build a small blimp, partly by contacting people who'd built small blimps.

Bill Arras found that Bill Watson still had the WD and it was just plain going to waste. Bill Watson figured Bill Arras, being World Champion Hot Air Balloonist, would make sure the WD would be used for the forces of good, and Gallagher didn't want to have anything more to do with it, so it was brought to Oregon. It sat in it's custom trailer while Bill and Reed discovered one reason why there aren't a lot more little blimps: We couldn't find a place tall enough to fit it. Eventually, Bill found a hangar with a 25 foot high door at the Madras Airport, which is a delight, because the airport is large and largely unused. And the flight service station owner Don Mobley thought the blimp was really great, made room for it in one of his hangars, and generally provided a lot of support. So Bill and Reed moved the WD to Madras, Oregon, assembled and inflated it Sept. 30, 2000, and first flew it Oct. 3.

It actually worked. It's beautiful construction is less important than the fact that it does what it was supposed to do, a rarity in small airships. Which is to putter around very slowly with precise control, only in very light winds.

Bill Arras added a couple of string trimmer 19cc engines with model airplane props. With a little peddling to help, he average 10 MPH for an hour. When he peddled hard in addition to full throttle, he got up to about 15 MPH, and then the envelope went unstable, as predicted. Of course, the consequences of "going un-stable" in the WD are a lot less significant than for most aircraft: the nose wandered a bit and he slowed down.

Check out our web site for some pictures courtesy of Richard Tetz, of himself, the engine setup, and Bill again exploring Willow Creek canyon. (Three 200K jpg's)

<< http://www.teleport.com/~reedg/whitedwarf.html >>

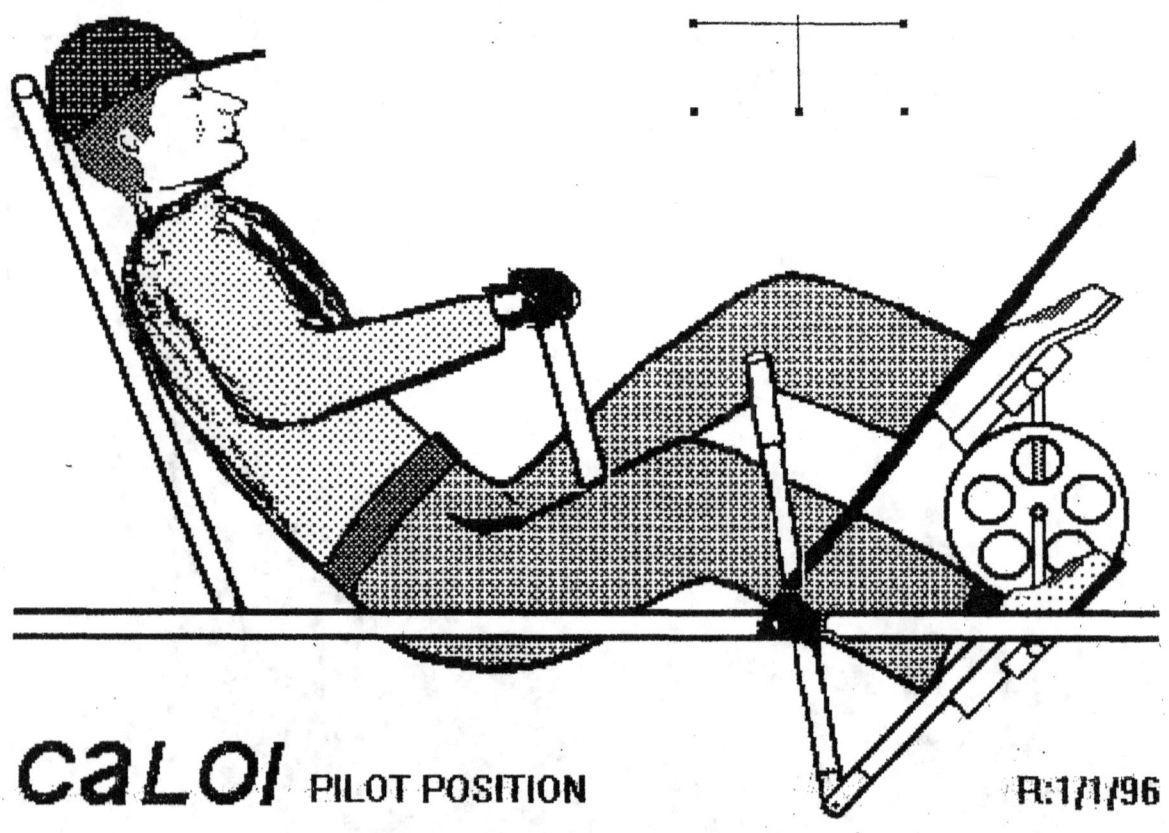

CALOI PILOT POSITION R:1/1/96

BOBIC DIRIGIBLE
(BOlha BIo Cinéca)
Sponsored by CALOI Bicycles Co.

<u>Envelope Dimensions:</u>
Volume = 300 cubic meters   (9711 cu.ft)
Gross Lift = 300 kilograms   (728 pounds)
Length = 17.5 meters   (57.4 feet)
Diameter = 6 meters   (19.7 feet)
Clearance height = 9 meters   (29.5 feet)
Materials = 300 sq.meters coated Polyester fabric
<u>Gondola Dimensions:</u>
Height (tower clearance)   (?? feet)
Length = 8.66 meters   (28.4 feet)
Width = 0.47 meters   (18.5 inches)
Side Clearance = 1.41 meters (4.62 feet)
Materials = High-strength aluminum tubing
<u>Propulsion system:</u>
Power; Human-pedal driven propeller
  chain-driven right-angle gearbox
Prop = Constant cord / zero twist
  design velocity = 15 kph   (9.3 mph)
Prop Materials; Hollow formed Balsa wood

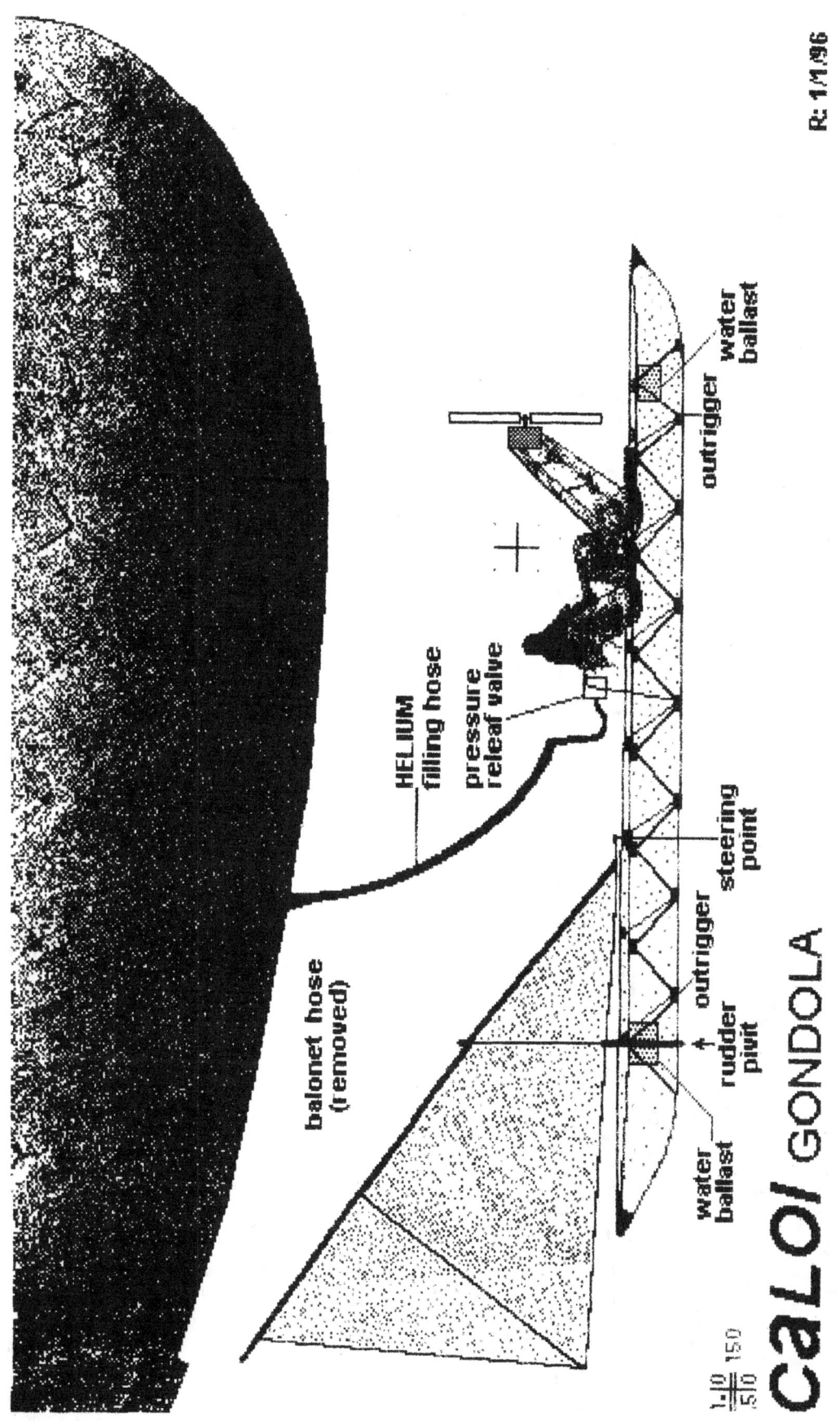

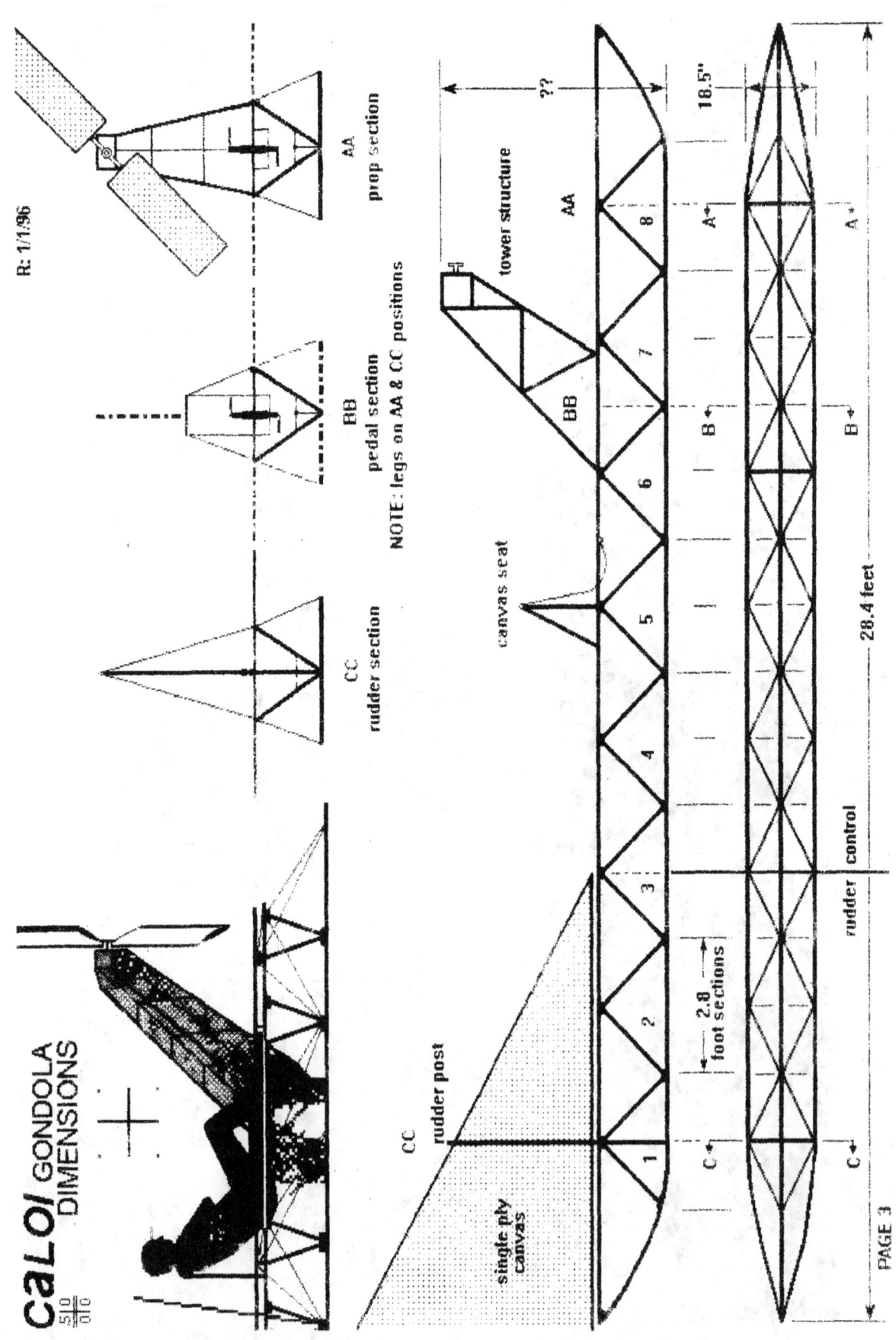

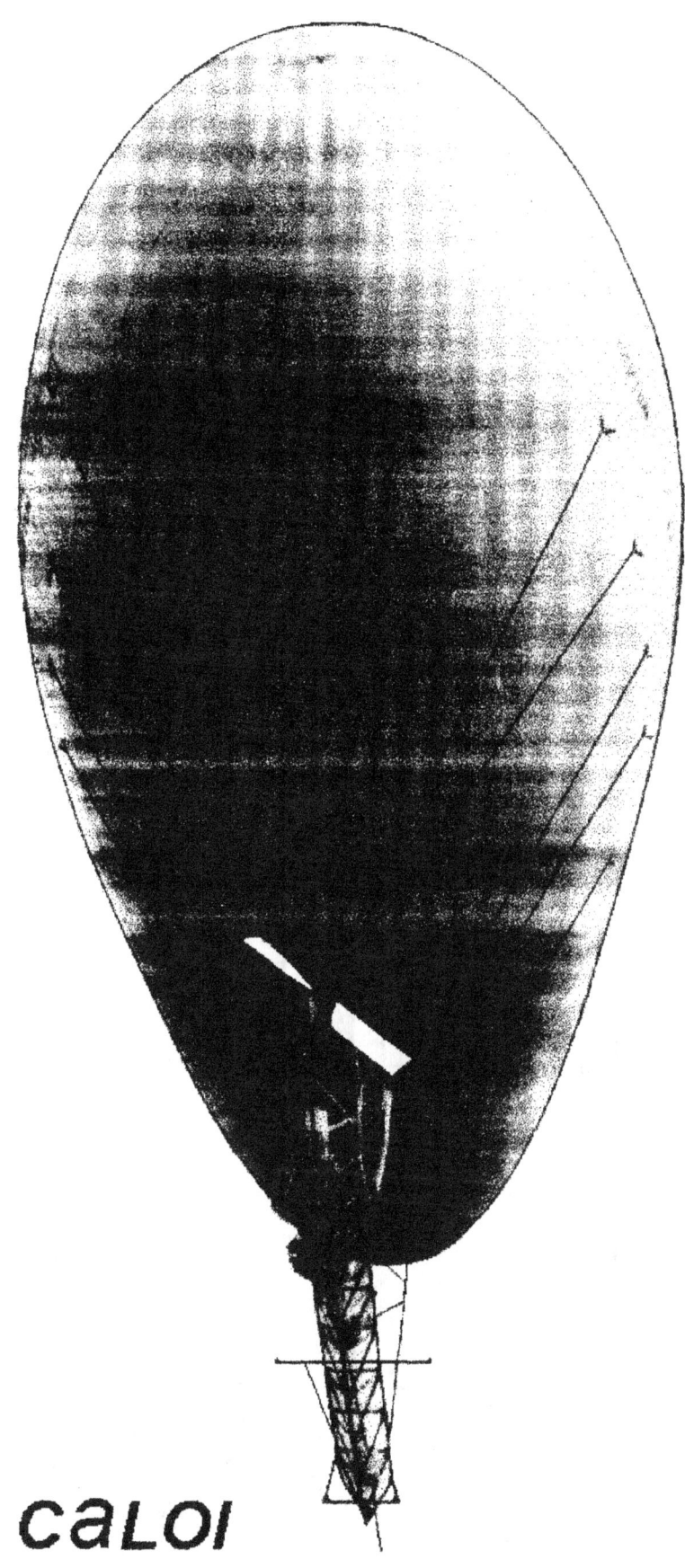

MUSCLE POWERED BLIMPS

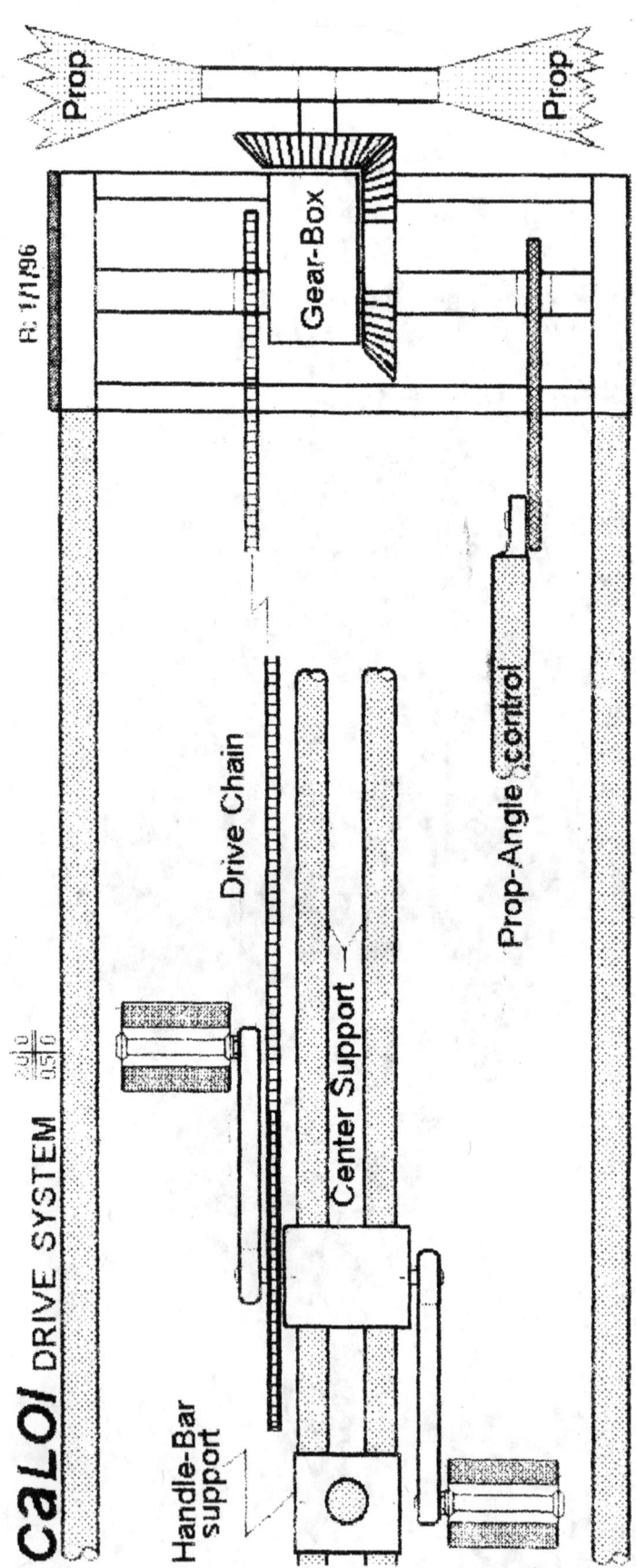

MUSCLE POWERED BLIMPS

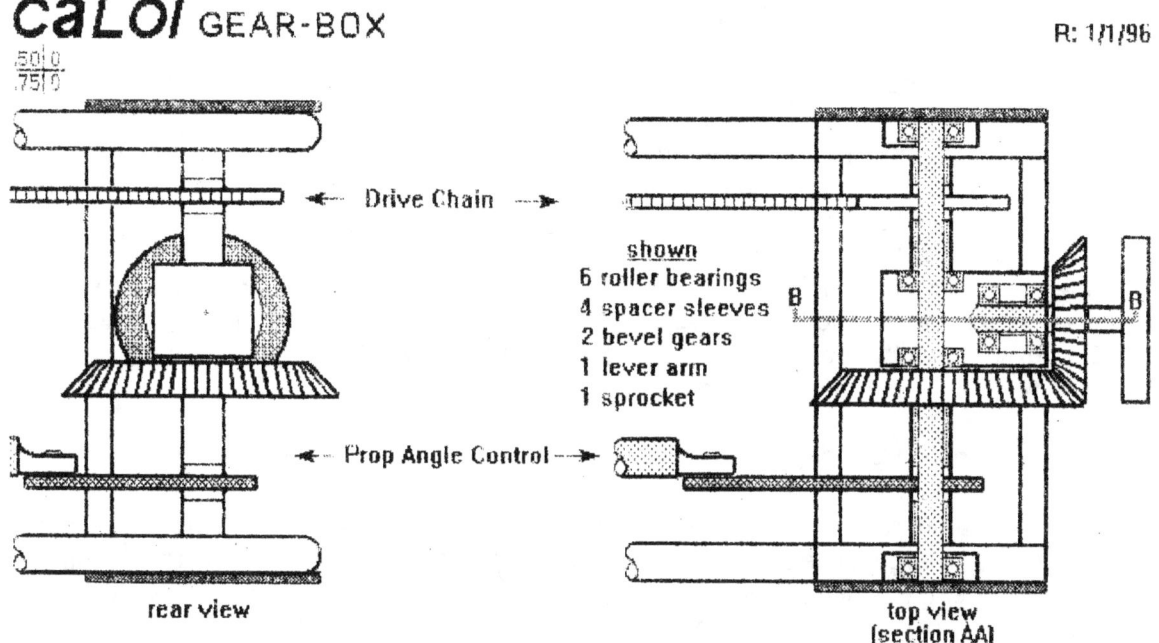

Caloi GEAR-BOX

R: 1/1/96

shown
- 6 roller bearings
- 4 spacer sleeves
- 2 bevel gears
- 1 lever arm
- 1 sprocket

rear view

top view (section AA)

NOTE: drawing not to scale

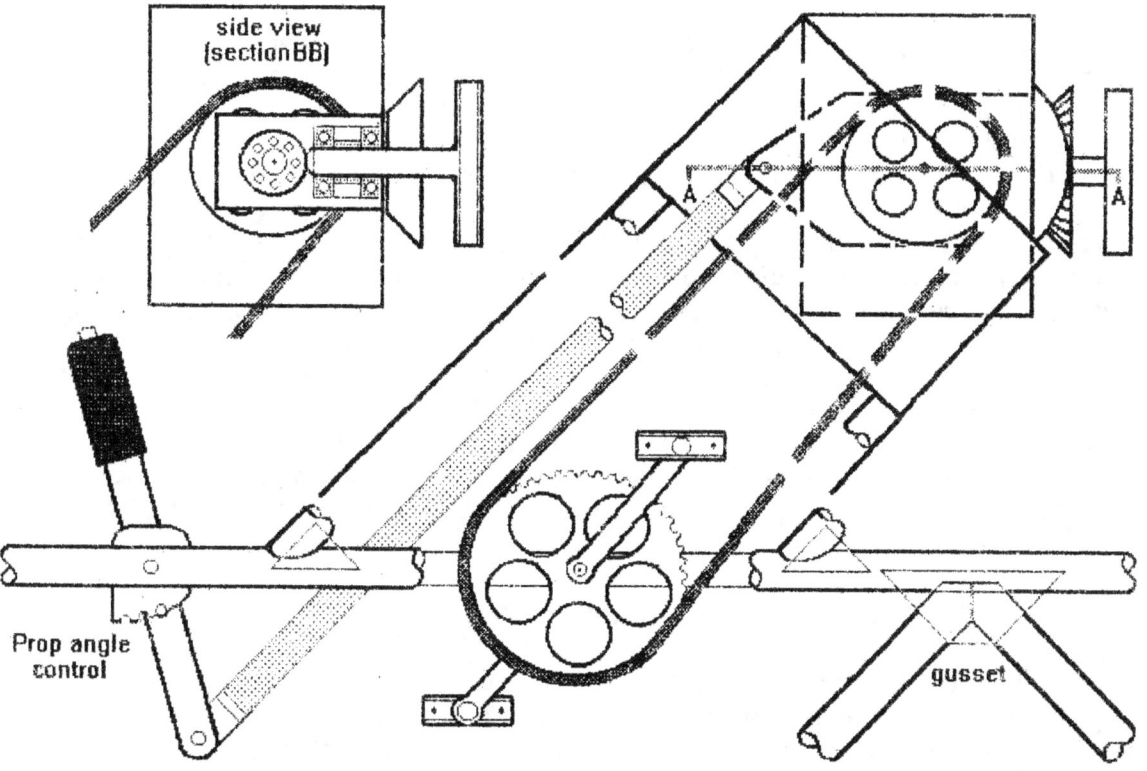

side view (section BB)

Prop angle control

gusset

BUILDING AIRSHIPS
(Submitted 1909 to Popular Mechanics Magazine)
By Glen Hammond Curtis, President
G. H. Curtis Mfg. Co. Hammondsport, N.Y.

Defining the Types of Flying Machines

The term airship, generally speaking, is applied to dirigible balloons, while the heavier-than-air classes are more commonly spoken of as flying-machines. The flying-machine, or aeronef, is divided into three classes; Aeroplanes, which consist of one or more horizontal planes designed to soar into the air by being propelled forward at an incline; the helicopter, in which the ascensive force is secured by the use of vertical screws, or propellers; and the ornithopter, or wing-beating machine. In this chapter we will endeavor to describe the most simple and practical form of dirigible balloons, while the other types will be taken up in successive issues.

How to Estimate Capacity for Lifting

The ascensive power of a dirigible or other balloon is secured by the use of gas contained in a large envelope of oiled fabric or rubber-coated cloth. For the dirigible balloon where the bulk of the gas bag is an important consideration, the use of hydrogen gas, which h as the greatest ascensive power, is desirable; 1,000 cu. ft. of hydrogen gas will lift about 65 lb.

How to Build a Dirigible Balloon

In building an airship, it is well to first determine the weight of the frame, propellers, engine, controlling mechanism and operator; then build, or purchase, the gas bag, of proper dimensions and sufficient capacity to lift the desired weight, together with a reasonable amount of ballast, which in a one-man outfit should be about 50 lb. Experience has taught us that a 7-hp, engine driving a suitable propeller will furnish sufficient pull to drive a one-man airship as fast as it can be readily controlled. Taking this as a basis, let us proceed to enumerate what our requirements and their respective weights will be: An engine of this power in the form of a 2-cylinder air-cooled motor will be the most desirable. This will weigh, with all appurtenances, about 100 lb., the engine alone only weighing 50 lb. From this it will be seen that in securing an engine we must not consider simply the catalog weight of the engine, which seldom includes the ignition system, oil or gasoline tanks, mufflers, etc. Placing the weight of an operator at 175 lb., frame 60 lb., propeller and shaft 40 lb., rudder, drag rope and ballast 100 lb., we have a total weight of 375 lb. Add 80 lb. for the weight of the gas bag and its suspension, and we have 455 lb., which divided by 65 gives us 7,000 cu. ft. of gas required to lift the machine.

Determining the Shape of the Balloon and Frame

It is apparent that an elongated balloon will pass through the air with much less resistance than a spherical balloon. We will, therefore, adopt a form in which the length is about four times the diameter. A diameter of 15 ft. and a total length of 60 feet gives us the desired cubic capacity. With this in mind, it is evident that a frame of considerable length must be constructed in order to support this long gas bag for the greater part of its length. This frame should be 40 ft. in length and can be constructed of spruce in the form of a triangle and properly guyed with wires at a weight of about 1-1/2 lb. per foot. The illustration, Fig. 1, shows the proper form and method of construction. The frame should be in the form of a triangle measuring about 3 ft. on a side. The length of the spruce sticks would be approximately 16 ft. These sections can then be butted and spliced together by short pieces, lapped on underneath, and fastened by bolting and lashing. Lashing is one of the best ways of making the joint of two or more pieces on an airship.

The frame is hung underneath the balloon by attaching it to the netting on either side by light-weight linen cords, as shown in the sketch. The longitudinal sticks of the frame should be triangular in form, while the cross sticks should be square. The engine should be mounted at about one-third the length of the frame from the forward end, and the power transmitted to the propeller shaft by the use of a heavy bicycle or motor-cycle chain. The propeller shaft should be made of 1-1/4-in. 16-gauge tubing, sup-ported by about five bearings and fitted with a thrust device for taking the pull of the propeller.

The propeller, Fig. 2, would be 10 feet in diameter, with an equal pitch. The pitch can be secured by fitting the braces which hold the arms of the propeller to the shaft at an angle of 20 degrees from each other. The arms for the propeller should be made of hickory or ash, and the canvas covering tacked on over light 1/4 in. by 1 1/2-in. slats mortised into these arms. The blade at its widest place should be 2 ft. For convenience in removing or replacing, in case of accident, the propeller should be made up separately from the shaft and attached to the shaft by two 1/4 inch tapered pins. The rudder should contain about 50 sq. ft. of surface and be braced in the manner shown in the sketch. It is preferable to cover this rudder with silk, the woodwork being of bamboo. The rudder, Fig. 3, is controlled by an endless cord running through a pulley in front of the operator, so that he can get hold of it with either hand.

Construction of the Gas Bag and Propeller

To build the gas bag is perhaps the most difficult part of the construction and requires the most skill. The builder must, after determining the size of his bag, divide it into'three sections: the forward taper, the straight cylinder, and the rear taper. The cylinder is composed of straight strips of equal length, while for the tapered ends the silk must be cut in the form of a triangle, with the sides cut on a slight curve. This can be secured by hanging the silk on the wall and attaching silk cord from end to end, marking the silk as the cord hangs. This will give a good form. All of the seams run lengthwise. There is no strain on the silk when inflated in a properly fitting net. While cotton fabric may be used, silk is by far the better. If properly treated it will last indefinitely. It is much stronger and lighter than other fabric. The first operation in building the balloon is to oil the material. The fabric should be cut in lengths and treated with linseed oil. The oil can best be applied by dipping in a large vessel or tub. The strips should then be hung by one end in a large room, of moderate temperature: This first coat of oil should dry in three or four days, although in some climates it takes considerably longer. After the strips of silk are given one coat of oil and thoroughly dried they may be cut to the proper shape to form the cylinder and cone-shaped ends of the gas bag.

The seams should run lengthwise, and each lap should be double stitched. After all of the silk has been sewed up, a manhole of about 15 in. in diameter should be made in the center, and a small neck, 6 in. in diameter, a little to the rear of the manhole, fitted for inflation. The balloon should then be blown up with air from a centrifugal blower, and a coat of oil put on by brush from the outside. A strip should be oiled the entire length

MUSCLE POWERED BLIMPS

and the balloon rolled over slowly and another strip oiled, etc., until the entire surface is covered. It may then be turned inside out, through the manhole, and the process repeated. After it has been given a sufficient number of coats to make it airtight, powdered soapstone, or French chalk, should be sprinkled over the entire surface inside and out, to prevent sticking. Care must be taken that expansion shall not occur from change of temperature and burst the bag.

The framework is suspended from the balloon by linen cords attached to a square mesh of Irish linen net, as shown in the illustration, Fig. 1. The net should be carefully adjusted over the bag before filling is commenced. After the balloon is fully inflated, the framework may be placed beneath it and the suspension cords attached, the rudder and propellers fitted on, and the machine is ready for a flight.

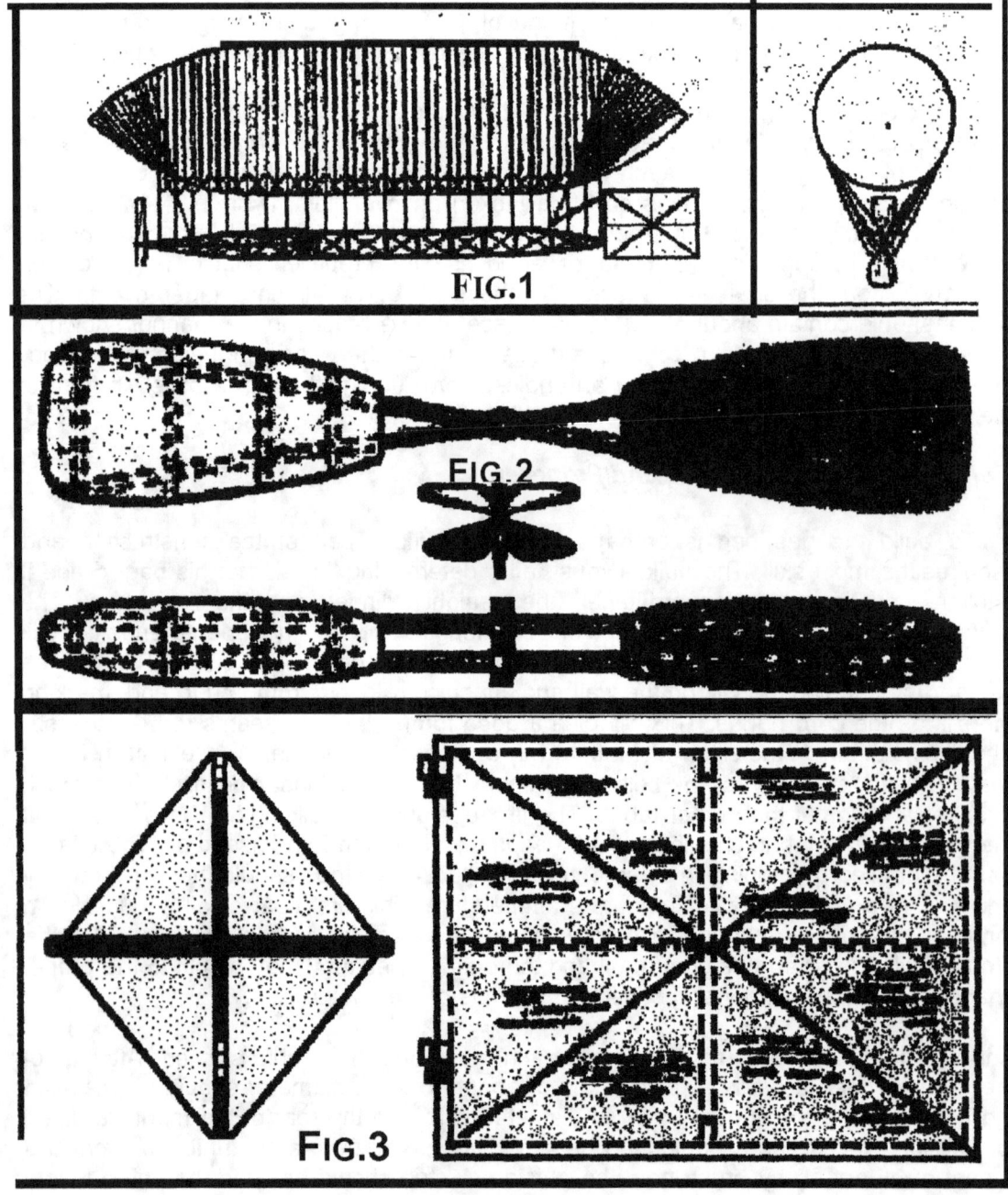

C. A. NICHOLS, JR.
CHILI, N.Y.

DEAR FRIEND:

Your answer to my advertisement received, and I take great pleasure in sending you herewith full particulars of the airship.

We have passed through the steam, steel and electric age, and are entering upon a new era, which presents greater possibilities than any we have passed through. The aerial age is upon us. It has only been a few years since the steam car, telegraph, telephone, electric light and the automobile were laughed at and considered useless devices invented by some freak mechanic, but time soon forced their true worth upon us.

Aeronautics is in its youth, but the developments of the last few years lead us to believe that before many years the air WILL be our chief source of transportation. Recent government contracts for air ships show the trend of development and it is no idle prophecy that the future fates of nations will be decided in the sky.

An airship ascension is always looked forward to with great interest but was considered an expensive and dangerous undertaking, and must be managed only by expert aeronauts. My system eliminates all objectionable features and makes it possible for any one without previous experience to give an airship demonstration.

My proposition is a profitable one because any merchant will pay you well to have his advertisement on the airship for the excitement, talk and comment that will bring him a profit many times over what it may cost him. Then again, many places feel that they cannot afford to hire a professional aeronaut for such times as the Fourth of July and other celebrations, yet can easily raise ten or fifteen dollars for an ascension of this kind. Until my system was given to the public aerial experimentation was a dangerous undertaking and required much capital but now with a small outlay you may be able to study aerial navigation and be among the first in this new field. Furthermore my system can establish you in a well paying business with smallest possible expense.

Wishing you success, I remain,

Very truly yours;

CHARLES A. NICHOLS, JR.

1 June 1921

SMALL Bicycle Parts Catalog

Consumer Edition

U.S. $5.00 suggested

A catalog of repair parts for quality components. Allows you to find the small repair parts you need so that you can ask your dealer for them by number.

•NOTE•
Some parts listed are not currently available. Please consult with your dealer.

Includes assembly schematics and product specifications

SHIMANO

Reprinted December 1993

Recycled

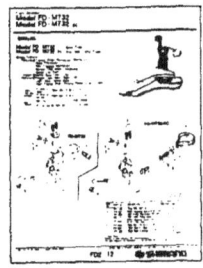
The complete alphanumeric code at the bottom of the page denotes the component type, its location in the catalog, and the sequence in which it was added to the catalog. For example, the page for the Deore XT-II Front Derailleur indexed as FD 2-12. This code represents a Front Derailleur component page (FD) in the second section of the catalog (2) which is the twelfth addition to that section (12).

Catalog Page Format

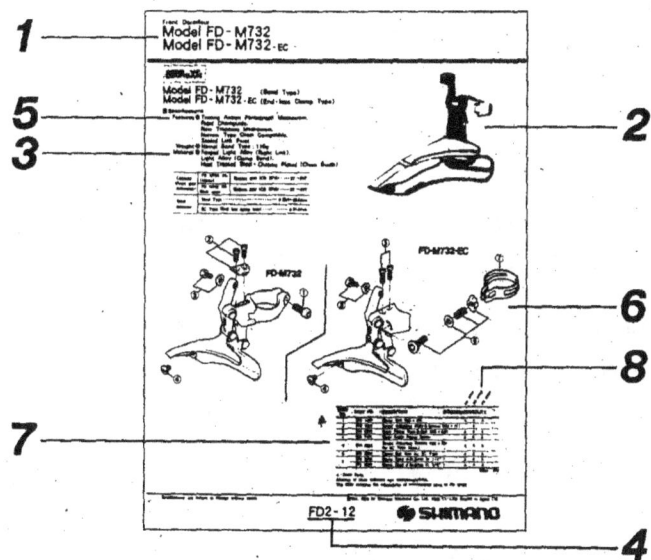

1. Component Model Number
The component's name and model number are printed at the top left of each page to allow easy referencing within the grouping.

2. Component Photo
A photograph of the component provides quick reference in regards to its appearance and visual verification.

3. Genenral Specifications
General component specifications are provided on the left side of the page. Especially useful when selling components separately, this data can be used to verify that component specs match the intended applications.

4. Index Code
The Index code printed at the bottom of the page aids group referencing and can also serve as a verification number when communicating with Shimano or a Shimano supplier who use the same catalog.

5. Features
The special features and functions that the component offers are included into specification sections.

6. Exploded Parts Drawing
The exploded parts drawing provides a detailed view of the construction and design of the component and simplifies the selection and ordering of replacement parts.

7. Part Names & Number
Used in conjunction with the exploded drawing, the component part name and number listing allows error-free ordering of replacement parts.

8. Interchangeability Chart
The right side of the listing contains a handy interchangeability chart which shows the extent to which the parts can be used in other Shimano components.

GLOSSARY

adjustable cup—the left-hand cup in a bottom bracket, used in adjusting the bottom bracket bearings and removed during bottom bracket overhaul

Allen key (Allen wrench)—a small L-shaped hexagonal wrench that fits inside the head of a bolt or screw

Allen wrench—see Allen key

all-terrain bike (ATB)—a bicycle with straight handlebars, sturdy fat tires, and wide-range gearing designed for off-road use

Alpine gearing—a gearing system in which a shift between chainwheels is equivalent to one-and-a-half shifts on the freewheel

binder bolt—the bolt used to fasten a stem inside a steerer tube or a seatpost inside a seat tube

bottom bracket—the cylindrical part of a bicycle frame that holds the crank axle, two sets of ball bearings, a fixed cup, and an adjustable cup

brake block—see brake pad

brake pad (brake block)—a block of rubberlike material fastened to the end of a brake caliper; it presses against the wheel rim when the brakes are applied

brake shoe—the metal part that holds a brake pad and is bolted to the end of a brake caliper

braze-ons—parts for mounting shift levers, derailleurs, water bottle cages, and racks, which are fastened to a bicycle frame through a type of soldering process known as brazing

butted tubing—tubing whose outside diameter remains constant but whose thickness is reduced in midsection where less strength is needed

cage—on a front derailleur, a pair of parallel plates that push the chain from side to side; on a rear derailleur, a set of plates in which pulleys are mounted to hold and guide the chain from cog to cog

calipers—brake arms that reach around the sides of a wheel to press brake pads against the wheel rim

cantilever brakes—rim brakes with pivoting arms mounted on fork blades or seatstays at or below rim level

chain—linked metal rope that connects the chainwheel to the back wheel, sized differently for different types of bikes (see derailleur chains)

chainring (chainwheel)—a sprocket attached to the right crankarm to drive the chain

chainring nut spanner—a special tool used to loosen the slotted nuts that fasten a chainring to a crankarm

chainstays—the two tubes of a bicycle frame that run from the bottom bracket back to the rear dropouts

chainwheel—see chainring

chain whip (chain wrench)—a tool consisting of a metal bar and two sections of chain, used in changing cogs on a freewheel

chain wrench—see chain whip

clincher tire—a tire whose edges hook under the curved-in hooked edge of a special rim, not commonly found anymore and often confused with the common wired-on tire (see wired-on tire)

cog—a sprocket attached directly to the rear hub on a single-speed bike and mounted on a freewheel on a multi-speed bike

cone—a bearing race that curves to the inside of a circle of ball bearings and works in conjunction with a cup

corncob—a term used to describe a cluster of cogs on a racing freewheel because of the small variation in number of teeth on adjacent cogs

cottered crank—a crankset in which the crankarms are fastened to the axle by means of threaded cotter pins and nuts

cotterless crank—a crankset in which the crankarms are fastened to the axle by means of nuts or bolts instead of cotter pins

crankarm—a part, one end of which is attached to the bottom bracket axle and the other holds a pedal, whose forward rotation provides the leverage needed to power the bicycle

crankarm fixing bolt—the bolt that holds a crankarm on the end of the axle in a cotterless crankset

crankset—a group of components that includes the bottom bracket removable parts, two crankarms, and one or more chainrings

crossover gearing—a gearing system whose shift sequence involves moving from the lowest to the midrange of gears on the smaller chainring, then crossing over to the larger chainring for the remainder of the gears

cross three—a spoking pattern in which a spoke passes over two and under a third spoke before being attached to the rim

C-spanner—a wrench whose end is shaped like a C, used to loosen the lockring on a bottom bracket

cup—the bearing race that curves around the outside of a ring of ball bearings and works in conjunction with a cone

derailleur—a lever-activated mechanism that pushes the chain off of one sprocket and onto another, thus changing the gear ratio

derailleur chain, narrow width—chain made especially for use on an "ultra" or narrow freewheel, often recognizable by bulging inner link plates and flush chain pins

derailleur chain, standard width—chain designed to fit a freewheel of standard width, usually characterized by straight-edged plates and chain pins that protrude slightly beyond the outer link plates

diamond frame—the traditional men's bicycle frame, the principal parts of which form a diamond shape

dish—offsetting of the hub in a rear wheel on a derailleur bike to make room for the freewheel and still allow the wheel to be centered within the frame

down tube—the tube running from the headset to the bottom bracket, one part of the main triangle on a bicycle frame

D-ring—a D-shaped ring found on many models of shift levers, used to adjust the level of tension on the inner parts of the lever

drop—the vertical distance from the horizontal line connecting the two wheel axles and the bottom bracket, one way of determining the location of the bottom bracket in relation to the rest of the bicycle frame

dropout—a slot in the frame into which the rear wheel axle fits (see fork tips)

dropout hanger—a threaded metal piece that extends below the right rear dropout, used as a mount for the rear derailleur

drops—the lower, straight portion of a turned-down-type handlebar set

dust cap—a metal cap that fits into a hub shell to keep contaminants out of hub bearings; a metal or plastic end cover for a spindle in a pedal or a cotterless crankset

face—to shave the outer edges of a bottom bracket shell or the upper and lower ends of a head tube to make them parallel with one another and square to the tube's centerline

fixed cup—the right-hand cup in a bottom bracket, ordinarily not loosened or removed during bottom bracket disassembly (see adjustable cup)

fixed gear—a cog attached to a hub without a freewheel; it always turns as fast as the wheel

fixed wheel—same as fixed gear, the kind of rear wheel found on track bikes

fixing bolt—a bolt used to hold a crankarm on an axle in a cotterless crankset

flange—the parts of a hub shell to which spokes are attached; also sometimes used to designate the circle of metal inside the teeth on a chainring

fork (front fork)—the part of the frame that fits inside the head tube and holds the front wheel; a term also sometimes applied to the part of the frame where chainstays and seatstays join to hold the rear axle

fork blades—the parallel curved tubes that hold the front wheel

fork crown—the horizontal piece on the upper part of the front fork to which the fork blades attach

fork rake (rake)—the shortest distance between the front axle and an imaginary line extending through the head tube downward toward the ground

fork tips—the slotted tips of the fork blades into which the front wheel axle fits

freewheel—a removable component attached to the rear hub on most types of bikes; it carries gear cogs on the outside and contains a ratcheting mechanism inside that allows the wheel to rotate forward while the pedals, chain, and gear sprockets remain still or move in reverse

front fork—see fork

front triangle (main triangle)—actually a quadrilateral with one short side, it is the section of a bicycle frame that consists of the head tube, the top tube, the seat tube, and the down tube

granny—colloquial term for the tiny inner chainring on a triple chainring crankset

half-step gearing—a gearing system in which a shift between chainrings in a double chainring set is equivalent to half a gear step on the freewheel

headset—the combination of cups, cones, and ball bearings that creates the bearing mechanism that allows the fork column to rotate inside the head tube

head tube—the shortest tube in the main triangle, the one inside of which the fork column rotates

hooks—the curved, dropped sections of a set of turned-down handlebars

hub—the center of a wheel consisting of a shell to which spokes attach and contains an axle along with two sets of bearings, bearing cones, lockwashers, locknuts, and parts for attaching the wheel to the frame

hub brake—any type of brake (disc, drum, or coaster) that operates through the wheel hub rather than the rim

idler pulley—the pulley in a rear derailleur that stays farthest from the freewheel cogs and functions to keep tension on the chain

jockey pulley—the pulley in a rear derailleur that stays closest to the freewheel cogs and guides the chain from cog to cog during a gear shift

knobby tires—heavy-duty tires with large rubber knobs spaced relatively far apart to provide traction in wet, muddy terrain

ladies' frame—the type of frame in which the top tube is replaced by a second down tube to make mounting and dismounting the bike easier

loaded tourer—a bicycle whose structure, geometry, and equipment is designed to allow a cyclist to travel with 40 or 50 pounds of gear

locknut—a nut used in conjunction with a washer or a second nut to lock a mechanism in place, such as the nut found at the upper end of a headset and in front of the calipers on many caliper brakes

lockring—the notched ring that fits on the left side of a bottom bracket and prevents the adjustable cup from turning

lockwasher—a washer with a small metal tang to prevent it from turning, such as the washer beneath the locknut on a headset or between the locknut and cone on a hub

lug—an external metal sleeve that holds two or more tubes together at the joints of a frame

main triangle—see front triangle

master link—a special link on a bicycle chain that can be opened by flexing a plate, removing a screw, or some other means besides driving out a rivet

mixte frame—a frame that replaces the top tube with twin lateral tubes that run all the way from the head tube back to the rear dropouts

mounting bolt—see pivot bolt

nipple—a small metal piece that fits through a wheel rim and is threaded inside to receive the end of a spoke

panniers—luggage bags used in pairs and fastened alongside one or both wheels of a bike

pick-up—see yoke

pin spanner—a wrench with pins on forked ends, used to turn an adjustable cup on a bottom bracket

pivot bolt (mounting bolt)—a bolt on which the arms of caliper brakes pivot and which also serves as the means for mounting the brakes on the bike frame

plain gauge tubing—tubing whose thickness remains constant over its entire length

Presta valve—a bicycle tube valve whose stem has a small nut on top, which must be loosened during inflation, instead of a spring such as is found on the Schrader valve

quick-release—a cam-lever mechanism used to rapidly tighten or loosen a wheel on a bike frame, a seatpost in a seat tube, or a brake cable within cable housing

quick-release skewer—a thin rod that runs through the center of a wheel axle; a cam-lever is attached to one end and the other end is threaded to receive a nut

quill—similar to the rattrap type of pedal except that the two sides of the pedal frame are joined by a piece of metal that loops around the dust cap

races—curved metal surfaces of cups and cones that ball bearings contact as they roll
rake—see fork rake
rattrap—the type of pedals that have thin metal plates with jagged edges running parallel on each side of the pedal spindle
rear triangle—a frame triangle formed by the chainstays, seatstays, and the seat tube
rim—the metal or wooden hoop of a wheel that holds the tire and tube and the outer ends of the spokes
rim brake—any type of brake that slows or stops a wheel by pressing its pads against the sides of the wheel rim
rollers—a stationary training device that consists of a boxlike frame and three rotating cylinders (one for a bike's front wheel and two for its rear wheel) on which the bicycle is balanced and ridden

saddle—seat on a bicycle; metal piece on a centerpull brake (see yoke)
Schrader valve—a tire valve similar to the type found on automobile tires
sealed bearings—bearings fastened in sealed containers to keep out contaminants
seamed tubing—tubing made from steel strip stock that is curved until its edges meet, then welded together
seamless tubing—tubing made from solid blocks of steel that are pierced and drawn into tube shape
seat cluster—the conjunction of top tube, seat tube, and seatstays near the top of the seat tube
seatpost—part to which the saddle clamps and which runs down inside the seat tube
seatstays—parallel tubes that run from the top of the seat tube back to the rear axle
seat tube—the tube that runs from just below the saddle down to the bottom bracket
sew-up tire—a tire with an inner tube stitched inside the casing; also known as a tubular
shallow angles—angles that position frame tubes relatively farther from vertical and closer to horizontal than do steep angles
spanner—another name for a wrench, applied to many bicycle tools
spider—the multiarmed piece to which the chainwheels are bolted, usually welded to or part of the right crankarm
spindle—another term for an axle (such as a pedal axle or a bottom bracket axle)
spoke—one of several wires used to hold hub in the center of a wheel rim and to transfer the load from the perimeter of the wheel to the hub and on to the frame
sports tourer—a bicycle whose structure, geometry, and components are designed to make it a compromise between one suitable for racing and one suitable for loaded touring—good for general pleasure riding
sprocket—a disc bearing teeth for driving a chain, a general term that applies both to chainrings and to freewheel cogs
steep angles—angles that position frame tubes relatively closer to vertical than do shallow angles
steerer tube—the tube that forms the top of the fork and rotates inside the head tube
stirrup cable (straddle cable, transverse cable)—on centerpull brakes, a short cable, each end of which attaches to a brake arm and which is pulled up at the center to activate the brakes

straddle cable—see stirrup cable

tandem—a bicycle that provides seats, bars, and pedals for two or more riders, one behind the other
tap—to cut female threads inside a tube or opening; also the name of the tool that does the cutting
top tube—the horizontal tube that connects the seat tube with the head tube
touring triple—a triple chainring crankset designed to provide the wide range of gears needed for loaded bicycle touring
tourist—a cyclist who takes short or long excursions by bicycle, often carrying several panniers containing clothing and camping equipment
transverse cable—see stirrup cable
trials—a type of ATB cycling competition that tests riders not on speed but on ability to maintain balance while navigating a bicycle around and over numerous obstacles such as rocks, trees, and steep, slippery terrain
tubular tire—a type of tire that has a tube sewn up inside the casing, also known as a sew-up

ultra 7—a freewheel designed to allow seven cogs to fit into the space normally taken up by six
ultra 6—a freewheel designed to allow six cogs to fit into the space normally taken up by five

wheelbase—the distance between the front and the rear axles on a bicycle
wide-step gearing—a gearing system in which the step between the two chainrings is considerably greater than that found in most other systems
wind trainer—a training device consisting of a frame in which a bicycle is fastened for stationary riding and a fan that creates wind resistance to simulate actual road riding
wired-on tire—a tire with a wire bead edge that fits inside a trough-shaped rim; the type of tire often inaccurately referred to as a "clincher"

yoke (pick-up)—a triangular metal piece used to connect the main brake cable with the stirrup cable in a centerpull brake system (also known as a saddle)

MUSCLE POWERED BLIMPS

GEAR INCH CHART

Formula: $\dfrac{\text{Wheel diameter} \times \text{number of teeth on chainwheel}}{\text{Number of teeth on freewheel cog}} = \text{Gear Inches}$

CHAINWHEEL COG TEETH \ FREEWHEEL COG TEETH	34	33	32	31	30	29	28	27	26	25	24	23	22	21	20	19	18	17	16	15	14	13	12
56	44.5	45.8	47.3	48.8	50.4	52.1	54.0	56.0	58.1	60.4	63.0	65.7	68.7	72.0	75.6	79.6	84.0	88.9	94.5	100.8	108.0	116.3	126.0
55	43.7	45.0	46.4	47.9	49.5	51.2	53.0	55.0	57.1	59.4	61.8	64.5	67.5	70.7	74.2	78.1	82.5	87.3	92.8	99.0	106.0	114.2	123.8
54	42.9	44.2	45.6	47.0	48.6	50.2	52.0	54.0	56.0	58.3	60.7	63.3	66.2	69.4	72.9	76.7	81.0	85.8	91.1	97.2	104.1	112.1	121.5
53	42.1	43.4	44.7	46.2	47.7	49.3	51.1	53.0	55.0	57.2	59.6	62.2	65.0	68.1	71.5	75.3	79.5	84.1	89.4	95.4	102.2	110.0	119.3
52	41.3	42.5	43.9	45.3	46.8	48.4	50.1	52.0	54.0	56.2	58.5	61.0	63.8	66.9	70.2	73.9	78.0	82.6	87.8	93.6	100.3	108.0	117.0
51	40.5	41.7	43.0	44.4	45.9	47.4	49.2	51.0	53.0	55.1	57.4	59.9	62.6	65.6	68.8	72.5	76.5	81.0	86.1	91.8	98.4	105.3	114.8
50	39.7	40.9	42.2	43.5	45.0	46.5	48.2	50.0	51.9	54.0	56.3	58.7	61.4	64.3	67.5	71.0	75.0	79.4	84.4	90.0	96.4	103.8	112.5
49	38.9	40.1	41.3	42.7	44.1	45.6	47.2	49.0	50.9	52.9	55.1	57.5	60.1	63.0	66.2	69.6	73.5	77.8	82.7	88.2	94.5	101.7	110.3
48	38.1	39.3	40.5	41.8	43.2	44.6	46.3	48.0	49.9	51.8	54.0	56.3	58.9	61.7	64.8	68.2	72.0	76.2	81.0	86.4	92.6	99.6	108.0
47	37.3	38.5	39.7	40.9	42.3	43.7	45.3	47.0	48.8	50.8	52.9	55.2	57.6	60.4	63.4	66.8	70.5	74.6	79.3	84.6	90.6	97.6	105.8
46	36.5	37.6	38.8	40.1	41.4	42.8	44.4	46.0	47.8	49.7	51.8	54.0	56.5	59.1	62.1	65.4	69.0	73.1	77.6	82.8	88.7	95.5	103.5
45	35.7	36.8	38.0	39.2	40.5	41.8	43.4	45.0	46.7	48.6	50.7	52.8	55.2	57.9	60.8	64.0	67.5	71.5	76.0	81.0	86.7	93.4	101.3
44	34.9	36.0	37.1	38.3	39.6	40.9	42.4	44.0	45.7	47.5	49.5	51.6	54.0	56.6	59.4	62.5	66.0	69.9	74.3	79.2	84.9	91.3	99.0
43	34.1	35.2	36.3	37.5	38.7	40.0	41.4	43.0	44.6	46.4	48.3	50.4	52.8	55.2	58.1	61.1	64.4	68.2	72.5	77.4	82.9	89.3	96.8
42	33.4	34.4	35.4	36.6	37.8	39.1	40.5	42.0	43.6	45.3	47.2	49.3	51.5	54.0	56.7	59.6	63.0	66.7	70.8	75.6	81.0	87.2	94.5
41	32.6	33.5	34.6	35.7	36.9	38.1	39.5	41.0	42.4	44.2	46.1	48.1	50.3	52.7	55.3	58.2	61.5	65.1	69.1	73.8	79.0	85.1	92.3
40	31.8	32.7	33.8	34.8	36.0	37.2	38.6	40.0	41.5	43.2	45.0	47.0	49.1	51.4	54.0	56.8	60.0	63.5	67.5	72.0	77.1	83.0	90.0
39	31.0	31.9	32.9	34.0	35.1	36.3	37.6	39.0	40.5	42.1	43.9	45.8	47.9	50.1	52.6	55.4	58.5	61.9	65.8	70.2	75.2	81.0	87.8
38	30.2	31.1	32.1	33.1	34.2	35.3	36.6	38.0	39.4	41.0	42.7	44.6	46.6	48.8	51.3	54.0	57.0	60.3	64.1	68.4	73.2	78.9	85.5
37	29.4	30.3	31.2	32.2	33.3	34.4	35.6	37.0	38.4	40.0	41.6	43.4	45.8	47.5	50.0	52.5	55.5	58.7	62.4	66.6	71.3	76.8	83.3
36	28.6	29.5	30.4	31.4	32.4	33.5	34.7	36.0	37.3	38.8	40.5	42.2	44.1	46.2	48.6	51.1	54.0	57.1	60.7	64.8	69.4	74.7	81.0
35	27.8	28.6	29.5	30.5	31.5	32.5	33.7	35.0	36.3	37.8	39.3	41.0	42.9	45.0	47.2	49.7	52.5	55.5	59.0	63.0	67.5	71.9	78.8
34	27.0	27.8	28.7	29.6	30.6	31.6	32.7	34.0	35.3	36.7	38.2	39.9	41.7	43.7	45.9	48.3	51.0	54.0	57.3	61.2	65.5	70.6	76.5
33	26.2	27.0	27.8	28.7	29.7	30.7	31.8	33.0	34.2	35.6	37.1	38.7	40.5	42.4	44.5	46.8	49.5	52.4	55.6	59.4	63.6	68.5	74.3
32	25.4	28.2	27.0	27.8	28.8	29.7	30.8	32.0	33.2	34.5	36.0	37.5	39.2	41.1	43.2	45.4	48.0	50.8	54.0	57.6	61.7	66.4	72.0
31	24.6	25.4	26.2	27.0	27.9	28.8	29.8	31.0	32.1	33.4	34.8	36.4	38.0	39.8	41.8	44.0	46.5	49.2	52.3	55.8	59.7	64.3	69.8
30	23.8	24.5	25.3	26.1	27.0	27.9	28.9	30.0	31.1	32.4	33.7	35.2	36.8	38.5	40.5	42.6	45.0	47.6	50.6	54.0	57.7	62.3	67.5
29	22.3	23.7	24.5	25.3	26.1	27.0	28.0	29.0	30.1	31.3	32.6	34.0	35.5	37.2	39.1	41.2	43.5	46.0	48.9	52.2	55.9	60.2	65.3
28	22.0	22.9	23.6	24.4	25.2	26.0	27.0	28.0	29.0	30.2	31.5	32.8	34.3	36.0	37.8	39.7	42.0	44.4	47.2	50.4	54.0	58.1	63.0
27	21.4	22.1	22.8	23.5	24.3	25.1	26.0	27.0	28.0	29.2	30.4	31.7	33.1	34.7	36.5	38.4	40.5	42.9	45.6	48.6	52.1	56.0	60.8
26	20.6	21.3	21.9	22.6	23.4	24.2	25.1	26.0	27.0	28.0	29.3	30.5	31.9	33.4	35.1	36.9	39.0	41.3	43.9	46.8	50.1	54.0	58.5
25	19.9	20.5	21.1	21.8	22.5	23.3	24.1	25.0	26.0	27.0	28.1	29.3	30.7	32.1	33.8	35.5	37.5	39.7	42.2	45.0	48.2	51.9	56.3
24	19.1	19.6	20.3	20.9	21.6	22.3	23.1	24.0	24.9	25.9	27.0	28.2	29.5	30.9	32.4	34.1	36.0	38.1	40.5	43.2	46.3	49.8	54.0

SHIMANO

HG Cassette Sprocket (7-Speed / ac-Group / Chromica Gear)
CS-HG60-C7ac CS-HG50-C7ac

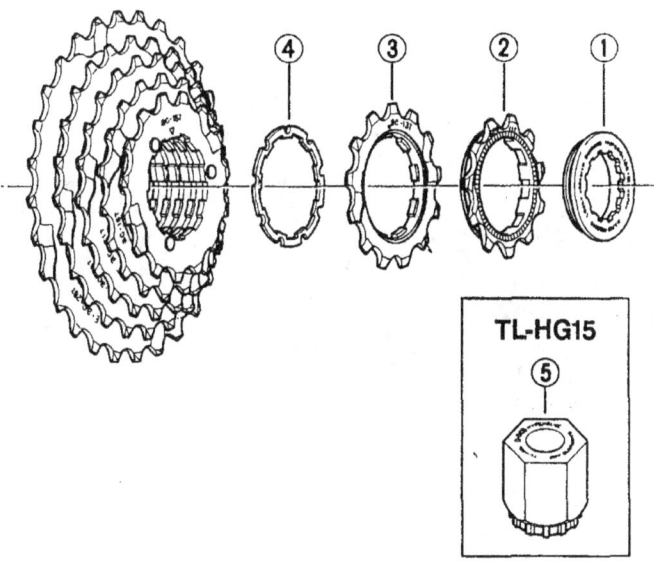

TL-HG15

- ac-group sprocket teeth combination: 11-13-15-18-21-24-28T
- Note: Use only the sprockets of ac-group together.

Q'TY	ITEM NO.	PART NO.	DESCRIPTION
	1	11E 0110	Lock Ring
	2	11F 1129-7	Sprocket Wheel 11T (Chromica / Built in spacer type)
	3	11F 1329-7	Sprocket Wheel 13T (Chromica / Built in spacer type)
	4	11E 0300	Sprocket Spacer (t=1mm / Black)
	5	120 0918	Lock Ring Removal Tool (TL-HG15)

CS4-27

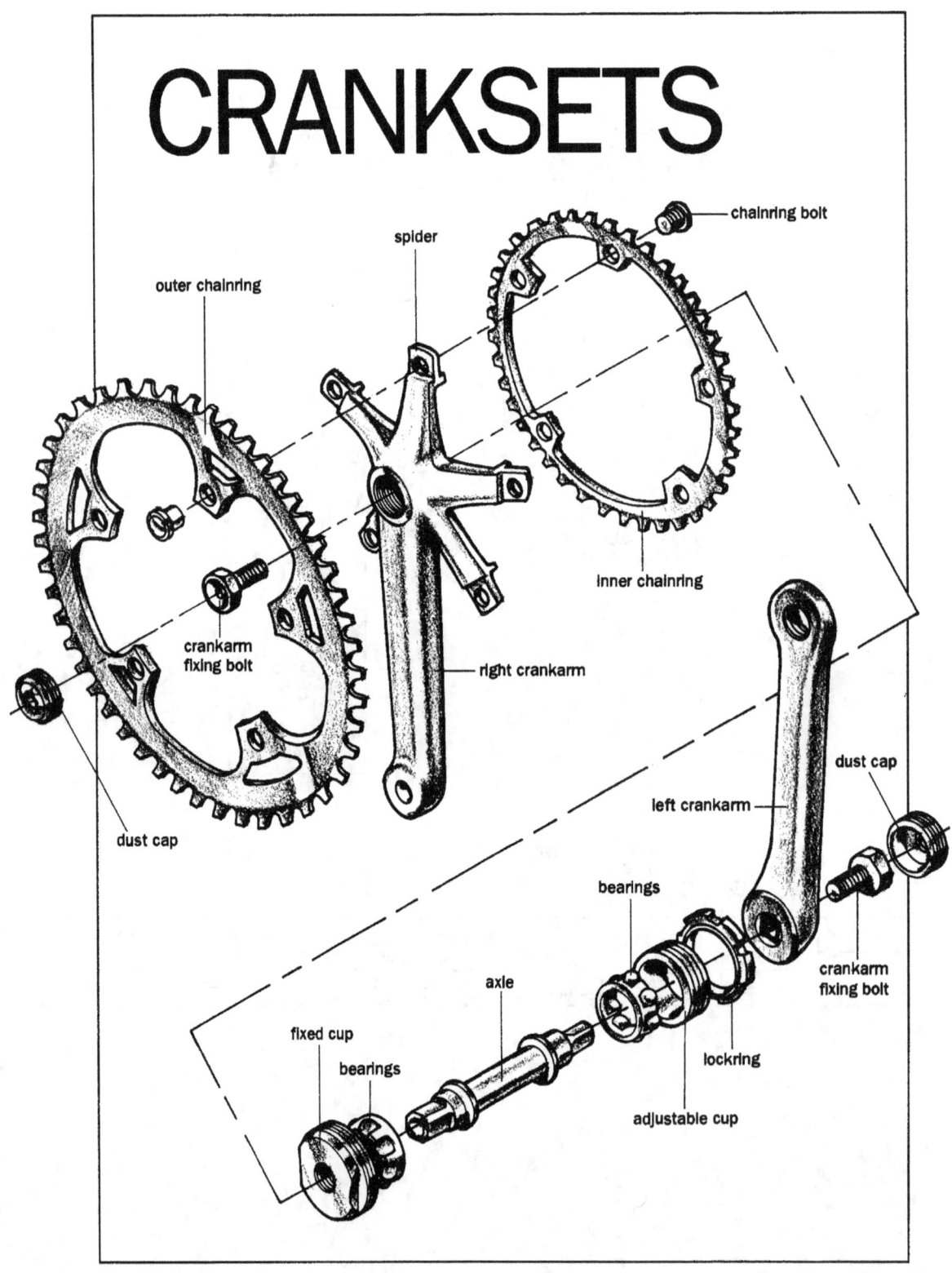

Front Chainwheel & Bottom Bracket Assembly
Model FC-6207
Model BB-6207

Model FC-6207
Model BB-6207 (B.B. Parts)

■ Specifications
- Weight ● 23.7oz.(672g.) 42T.~52T. 170mm
  11.5oz.(326g.) (B.B.Parts)
- Material ● Light Alloy · Anodized, Buff Finish
- Chain Ring ● 1/2" × 3/32" Chain
- Teeth ● FC-6207/Inner ; 39T.~45T.
  Outer : 48T.~53T.
- Pitch Circle Diameter ● 5-1/8" (130mm.)
- Crank Length ● 6-1/2" (165mm.), 6-3/4" (170mm.),
  6-25/32" (172.5mm.), 6-7/8" (175mm.)
- Crank Thread ● BC9/16 × 20T.P.I.(English)
- Cup Thread ● BC1.37 × 24T.P.I.(English)
  M35 × 1.0 (French), M36 × 24T.P.I.(Italian)
- B.B. Size ● 68-W-116 (D-3L), 70-W-119 (D-5LL)
- Material of Chain Ring ● Light Alloy · Anodized Finish
- Features ● Computer Aided Design (CAD), Sealed Mechanism
- Chain Line ● 15/16" (23.5mm.)

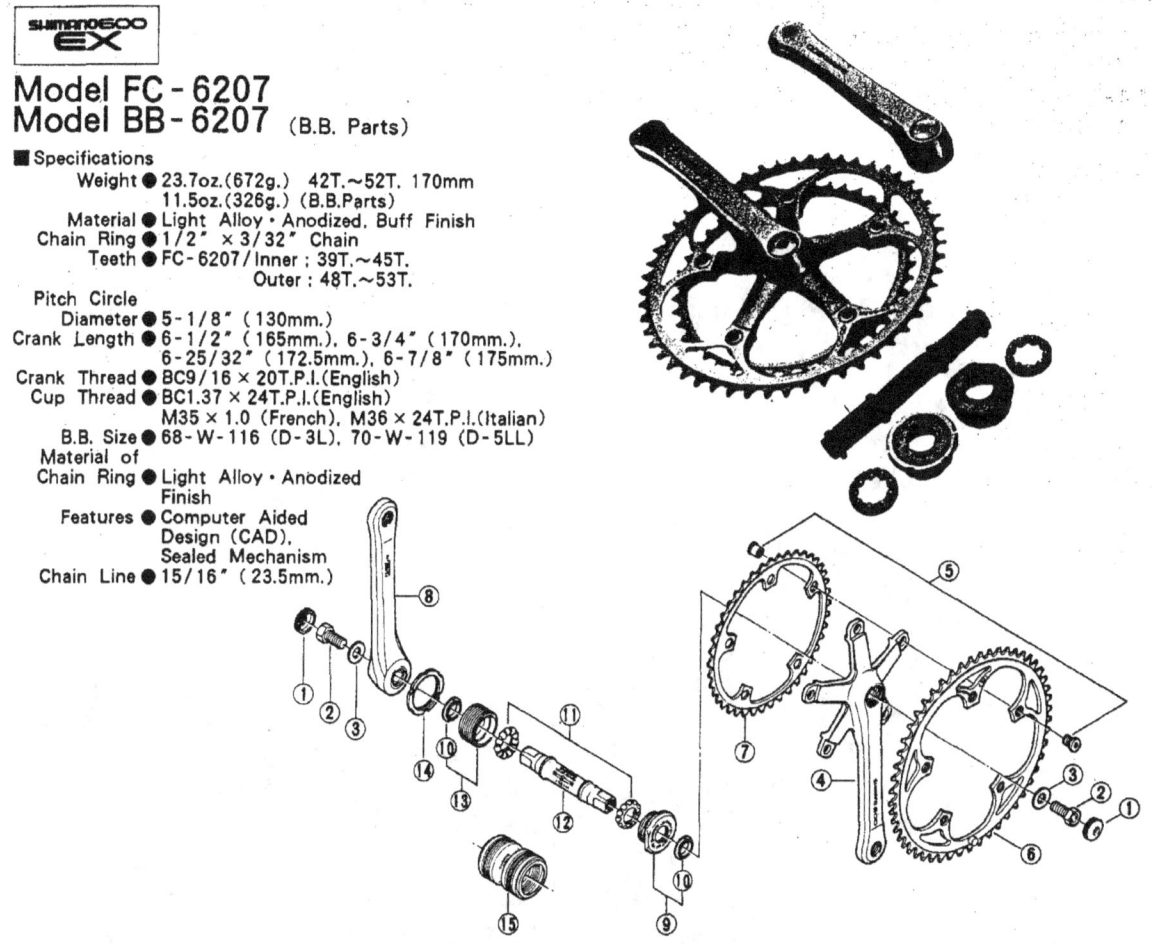

ITEM NO.	PART NO.	DESCRIPTION
1	144 2001	Crank Arm Cap
2	134 0300	Crank Arm Fixing Bolt
3	134 0400	Crank Arm Fixing Washer
4	144 2100	Right Crank Arm 165mm
	144 2300	Right Crank Arm 170mm
	144 3100	Right Crank Arm 172.5mm
	144 3200	Right Crank Arm 175mm
5	132 9801	Chainring Fixing Bolt & Nut (5sets)
6	144 4800	Chainring 48T
	144 4900	Chainring 49T
	144 5000	Chainring 50T
	144 5100	Chainring 51T
	144 5200	Chainring 52T
	144 5300	Chainring 53T
7	145 3910	Chainring 39T
	145 4010	Chainring 40T
	145 4110	Chainring 41T
	145 4210	Chainring 42T
	145 4310	Chainring 43T
	145 4410	Chainring 44T
	145 4510	Chainring 45T
	144 4650	Chainring 46T
8	144 2500	Left Crank Arm 165mm

ITEM NO.	PART NO.	DESCRIPTION
	144 2700	Left Crank Arm 170mm
8	144 3500	Left Crank Arm 172.5mm
	144 3600	Left Crank Arm 175m
9	179 9801	Right Hand Fixed Cup (1.37" × 24T) English Thread & Seal Cap
	179 9802	Right Hand Fixed Cup (35 × P1) French Thread & Seal Cap
	179 9803	Right Hand Fixed Cup (36 × 24T) Italian Thread & Seal Cap
10	179 0100	Seal Cap
11	189 9810	Ball Retainer (1/4") 2pcs.
12	148 0104	Spindle (D-3L) 116mm for Shell Width 68mm
	148 0113	Spindle (D-5LL) 119mm for Shell Width 70mm
13	179 9811	Left Hand Adjustable Cup (1.37" × 24T) English Thread & Seal Cap
	179 9812	Left Hand Adjustable Cup (35 × P1) French Thread & Seal Cap
	179 9813	Left Hand Adjustable Cup (36 × 24T) Italian Thread & Seal Cap
14	137 0500	Lock Ring (1.37" × 24T) English Thread
	137 0510	Lock Ring (35 × P1) French Thread
	137 0520	Lock Ring (36 × 24T) Italian Thread
15	179 2100	Dust Seal
16	179 11100	Complete English BB
	179 15400	Complete Italian BB

8607 - U603E

Specifications are subject to change without notice.   ©Mar. 1989 by Shimano Industrial Co., Ltd. 0389 C1/1.3M

FC7-6

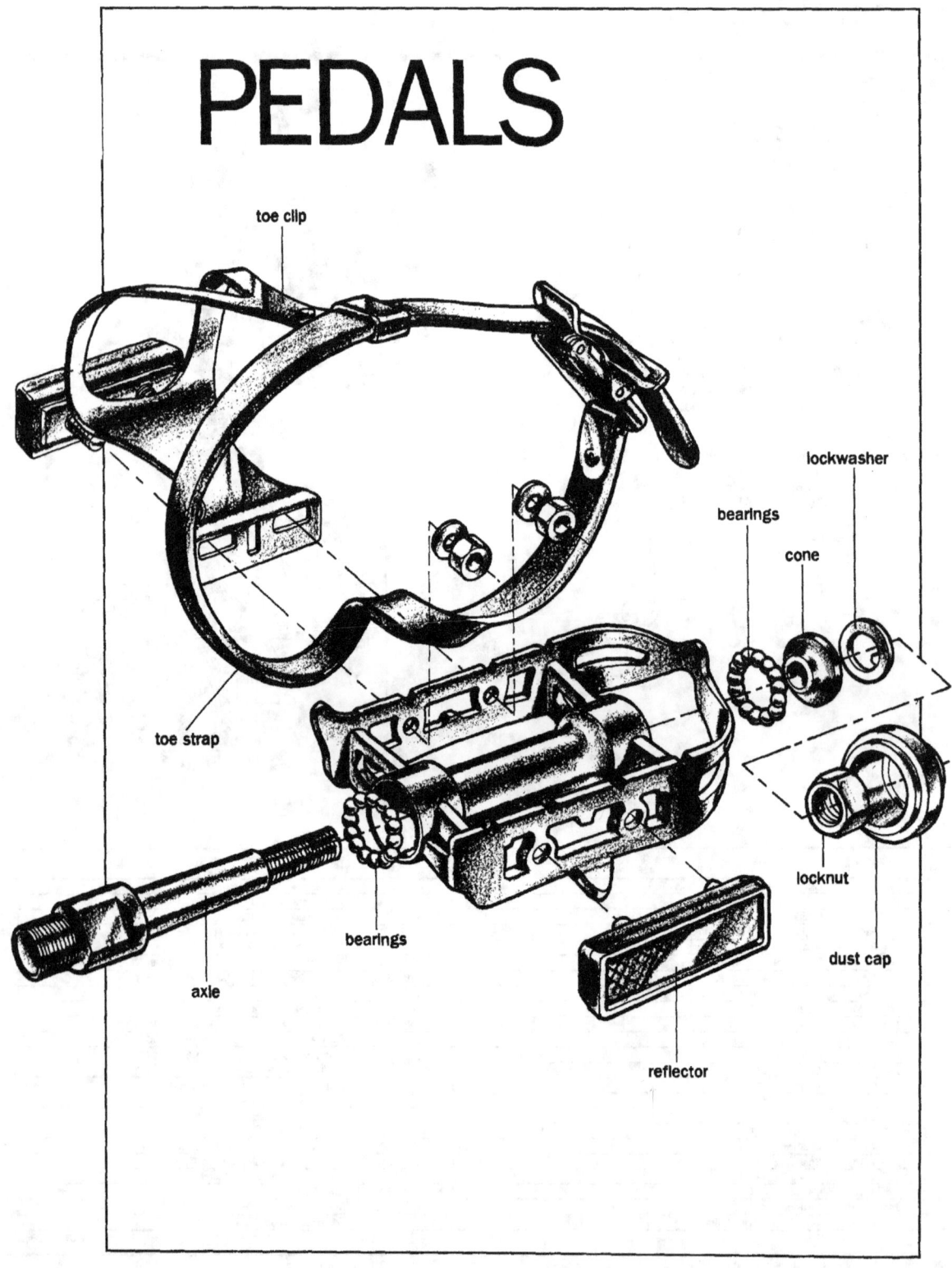

MUSCLE POWERED BLIMPS

Pedal
Model PD-7400

DURA-ACE
Model PD-7400

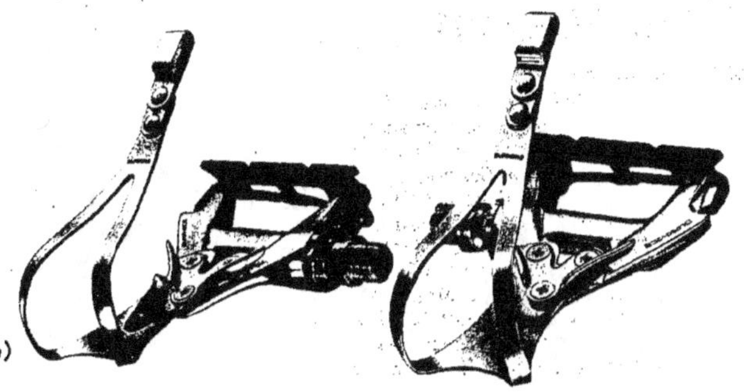

Specifications
- Features: Compound Structure of Needle Bearing and Cup & Cone, Aerodynamic Design, 34° Road Clearance, Sealed Mechanism
- Weight: Light Alloy : 13.1oz.(372g.) Pair. (Including Toe Clip, L-size)
- Material:
  - Light Alloy • Buff Finish (Body)
  - Light Alloy • Anodized Finish (Front Plate)
  - Light Alloy • Black Solid Anodized Finish or Steel • Chrome-Plated (Rear Plate)
  - Nickel Cr-Mo Steel • Chrome-Plated/Polished Finish (Axle)
  - Steel • Chrome-Plated or Light Alloy • Anodized Finish (Toe Clip)
- Toe-Clip Size: M, L, LL
- Crank Thread: English BC 9/16 × 20T.P.I.

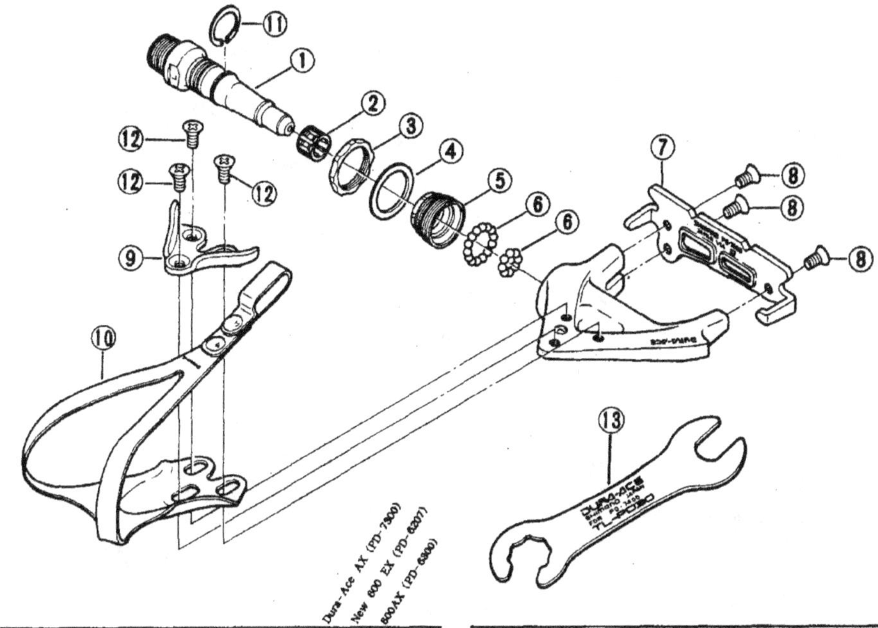

ITEM NO.	PART NO.	DESCRIPTION	INTERCHANGEABILITY Dura-Ace AX (PD-7300)	New 600 EX (PD-6207)	600AX (PD-6300)
1	440 9801	Pedal Axle (Right)			
	440 9803	Pedal Axle (Left)			
2	440 1000	Needle Bearing			
3	440 0900	Lock Nut			
4	440 0800	Seal Ring			
5	440 0700	Cup			
6	-	Steel Ball (1/8″) 22pcs.			
7	440 1210	Right Side Rear Plate (Light Alloy)	B		
	440 1310	Left Side Rear Plate (Light Alloy)	B		
	440 1200	Right Side Rear Plate (Steel)	B		
	440 1300	Left Side Rear Plate (Steel)	B		
8	440 1410	Plate Fixing Screw for Rear Plate	B	B	B
9	440 1110	Front Plate (Light Alloy)	B	B	B
	440 1100	Front Plate (Steel)	B	B	B
10	440 9010	Toe-Clip M-Size (Steel)	B	B	B
	440 9011	Toe-Clip L-Size (Steel)	B	B	B
	440 9012	Toe-Clip LL-Size (Steel)	B	B	B
	440 9013	Toe-Clip M-Size (Light Alloy)	B	B	B
	440 9014	Toe-Clip L-Size (Light Alloy)	B	B	B
	440 9015	Toe-Clip LL-Size (Light Alloy)	B	B	B
11	440 1800	Stop Ring			
12	440 1410	Plate Fixing Screw for Front Plate			
13	460 0903	Pedal Spanner (Model TL-PD30)			

B: Parts usable, but differ in materials, appearance, finish, size, etc.
The table indicates the adaptability of conventional parts to the PD-7400.

8703 - U676D

Specifications are subject to change without notice.   ©Mar. 1989 by Shimano Industrial Co., Ltd. 0389 C1/1.3M   CIS

MUSCLE POWERED BLIMPS

Pedal
Model PD-M730

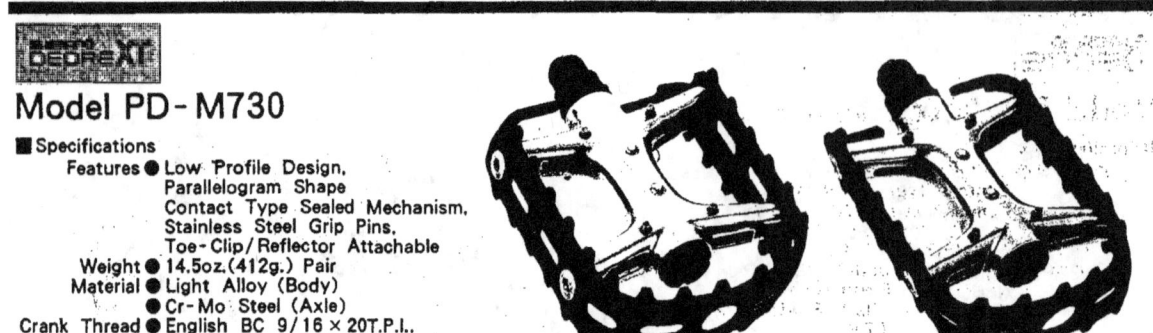

Model PD-M730
■ Specifications
Features ● Low Profile Design,
  Parallelogram Shape
  Contact Type Sealed Mechanism,
  Stainless Steel Grip Pins,
  Toe-Clip/Reflector Attachable
Weight ● 14.5oz.(412g.) Pair
Material ● Light Alloy (Body)
  ● Cr-Mo Steel (Axle)
Crank Thread ● English BC 9/16 × 20T.P.I.

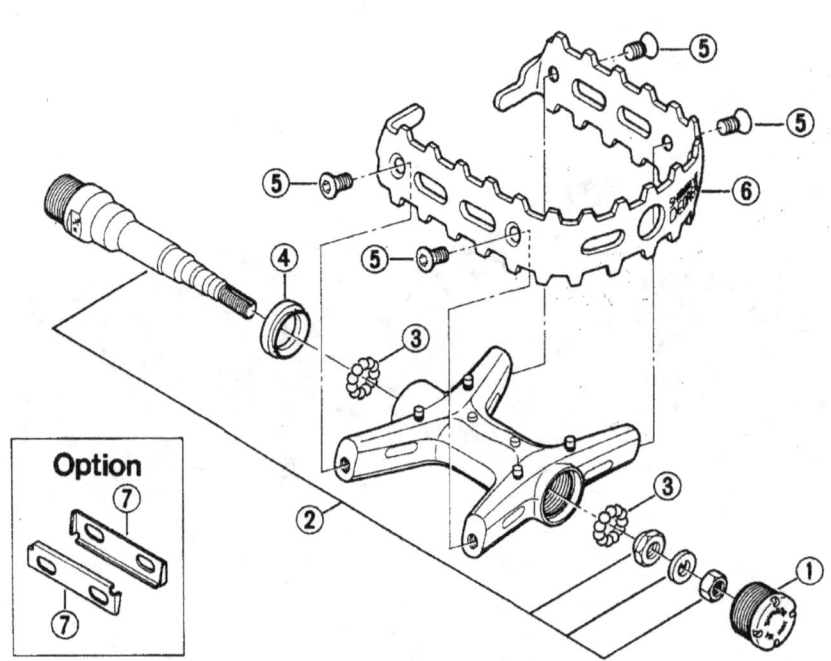

ITEM NO.	PART NO.	DESCRIPTION
1	416 0900	Cap
2	416 9801	Pedal Axle Unit (Right/B.C. 9/16″ × 20T.P.I.)
	416 9803	Pedal Axle Unit (Left/B.C. 9/16″ × 20T.P.I.)
3	—	Steel Ball (1/8″) 11pcs.
4	416 1200	Seal
5	416 1300	Plate Fixing Screw
6	416 0300	Right Side Plate
	416 0400	Left Side Plate
7	416 9810	Toe-Clip Adapter (2sets) Option

8702-U760D

Specifications are subject to change without notice.   ©Mar. 1989 by Shimano Industrial Co., Ltd. 0389 C1/1.3M

PD8-10

CHAINS

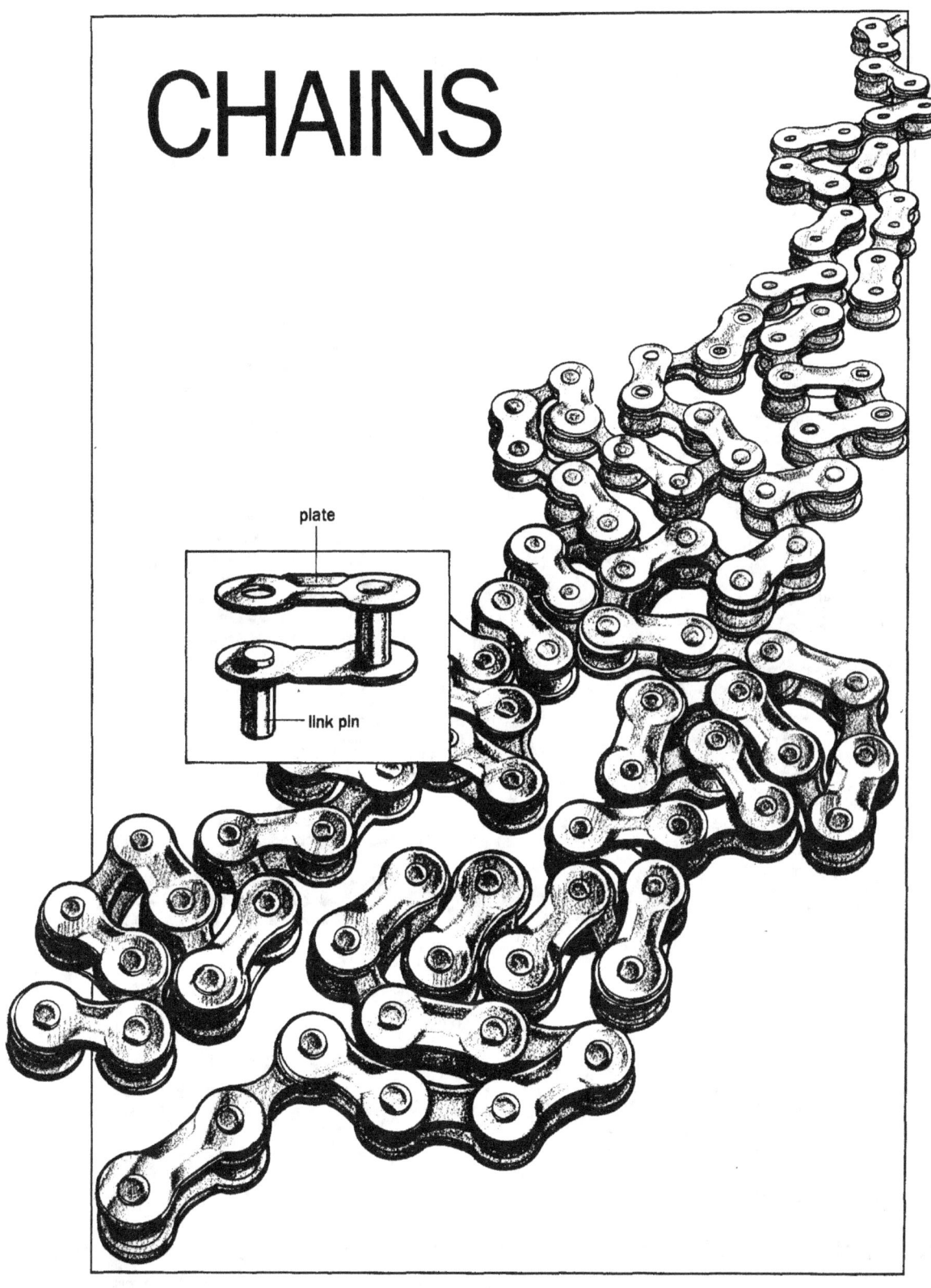

MUSCLE POWERED BLIMPS

Chain
Model CN-7400   Model CN-MT62   Model CN-UG30
Model CN-6208   Model CN-UG50
Model CN-M732   Model CN-6130

Model CN-7400
■ Specifications
- Narrow Type Uniglide Chain, Bushingless Design, Boron Finished Link Pins
- Size : 1/2" × 3/32" Chain
- Roller Link Plate : Nickel Finish, Pin Link Plate : Nickel Finish

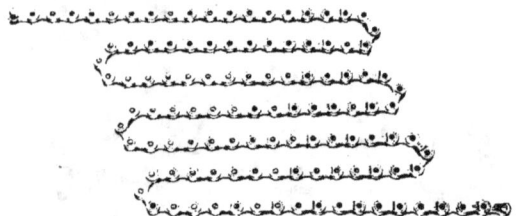

Model CN-6208 (Silver)
■ Specifications
- Narrow Type Uniglide Chain, Bushingless Design
- Steel
- Surface Treatment/ Roller Link Plate : Black Finish, Pin Link Plate : Silver Finish

Model CN-M732
■ Specifications
- Hyperglide Compatible
- Narrow Type Uniglide Chain, Bushingless Design
- Steel
- Roller Link Plate : Black Finish, Pin Link Plate : Silver Finish

Model CN-MT62
■ Specifications
- Hyperglide Compatible
- Narrow Type Uniglide Chain, Bushingless Design
- Steel

Model CN-UG50
■ Specifications
- Narrow Type Uniglide Chain.
- Black Finish
- Steel

Model CN-6130 (Silver or Gold)
■ Specifications
- Uniglide Chain
- Roller Link Plate : Black Finish
- Pin Link Plate : Silver or Gold Finish

Model CN-UG30 (Black)
■ Specifications
- Uniglide Chain
- Black Finish

Specifications are subject to change without notice.   © Mar. 1989 by Shimano Industrial Co., Ltd. 0389 CI/1.3M

CN15-1

RECOMMENDED LUBRICANTS

Part	Lubricant
Ball bearings	White lithium grease/bike grease
Bottom bracket axle	White lithium grease/bike grease (do not lubricate tapered ends)
Brake cable	Spray lubricant with Teflon (if lined housing) White lithium grease/bike grease (if unlined housing)
Brake pivot	Spray lubricant with Teflon or silicone
Brake spring	White lithium grease/bike grease
Chain	Oil with high water resistance, melted paraffin
Derailleur pivots	Spray lubricant with Teflon or silicone
Freewheel (inside)	Medium-weight oil (cycle oil or motor oil)
Hub with internal gears	Medium-weight oil (cycle oil or motor oil)
Seatpost (section inside seat tube)	White lithium grease/bike grease
Stem (section inside steerer tube)	White lithium grease/bike grease
Threads (bottom bracket, freewheel body, steerer tube, etc.)	White lithium grease, bike grease, or antiseize compound
Wheel axle	White lithium grease/bike grease

MUSCLE POWERED BLIMPS
Airship
KNOWLEDGEABLE CONTACTS

ALLEN, Brian
592 E. Poppyfields Drive
Altadena, CA 91001
JET PROPULSION LAB
holder of ALL p/p world records
Pilot, White Dwarf
626/794-0711(H
ballen@earthlink.net

BATES, Donald
2742 Swansboro Rd.
Placerville, CA 95967
PROP ENGINEERING
Has prop program for sale
prop maker
530/622-1886(W
www.innercite.com/~bateseng

CARMICHAEL, Bruce
34795 Camino Capistrano
Capistrano Bh., CA 92624
Aerodynamicist, ret.
Low speed design specialist
book writer
949/496-5191(5?)
TWITT member

GRILLMAIR, Carl
JPL Mail stop 100-22
Pasadena, CA 91125
JET PROPULSION LAB intern
Owner, White Dwarf (1998)
lives with B.Allen(?)
626/397-7065(H
carl@bugaboo.ipac.caltech.edu

JEX, Henry
1615 Georgina Ave.
Santa Monica, CA 90402
JEX ENTERPRISES
Consulting Aerodynamicist
Did a lot of work for McCreedy
310/394-2163
JexHR@aol.com

KICENIUK, Taras
1200 Bedford Street
Simi Valley, CA 93065
Low speed props
good/practical ultra-light acft.designer/pilot
has degenerative eyesight
805/525-8765(H
TWITT speaker

LARRABEE, Gene
1800 Knoxville Ave.
Long Beach, CA 90802
MIT Prof.of Aerodynamics, ret.
designer of White Dwarf prop
Suffered stroke early 1998
562/493-1195(H
TWITT speaker

McCREEDY, Paul
2222 Huntington Drive
Monrovia, CA
AEROVIRONMENT
Guru of acft.miniaturization
famous aircraft pioneer
626/357-9983(W
818/793-7429(H
TWITT speaker

McCOY, Chris
8608 Foothill Blvd.
Sunland, CA 91040
pedal-power blimp builder
friend of Allen & Grillmaire
818/353-0544(H
909/396-4261(F
cmcoy@sigmanet.com

RECKS, Robert
1148 Third Ave. #44
Chula Vista, CA 91911
THINK AIRSHIPS
airship builder/pilot/author
Avia.technical publications
619/427-5097(H
TWITT member
r.recks@juno.com

WATSON, Bill (Wil.C.)
Royal Ave.
Simi Valley, CA 93065
Radio control expert
builder, White Dwarf (has physical custody)
805/584-0975(W
805/526-4604(H

MUSCLE POWERED BLIMPS

BIBLIOGRAPHY

ARTICLES

ALLEN, Brian L.	*	Building & Flying a Human-Powered Airship		
		Sport Aviation Magazine, p21	August	1985
ALLEN, Brian L.	*	Pedal Powered Flight		
		AIAA Student Journal, p6	Summer	1988
CAMACHO, Zilda	*	Flight of the Brazilian Dirigible *CALOI*		
		Aviacao em Revista (Brazil), p12	October	1989
GROSS, Albert	*	Human-Powered Land Vehicles		
		Scientific American p142-52	December	1983
LARRABEE, Gene	*	The Screw Propeller		
		Scientific American p134-46	July	1980
LISSAMANN, P.B.S.		Flight Aerodynamics at Speeds under 5m/s		
		Aerovironment; Monrovia, CA		1980
MANTELL, Matt	*	Riis's Pieces (Scientific bicycling)		
		VeloNews p56?	July 27	1998
McMASTERS, John		An Analytical Survey of Low Speed Flying Devices		
		AIAA paper #74-1019		1974
PATIKY, Mark	*	White Dwarf & Pedal Pushers;		
		Air Progress, p	May	1981
WARNER, Bill	*	Man-Powered Blimps;		
		Model Aviation, p	May	1982

BOOKS

BURKE, James		Case Study in Aircraft Design	1980
		Aerovironment Monrovia, CA	
CROUCH, Tom	**	Eagle Aloft Smithsonian Press	1983
GLAUERT, Herman		Aerodynamic Theory TL570. D865	
		Verlag, Germany	1936
REAY, David		History of Man Powered Flight	
		Pergamon Press Oxford	1977
DWIGGINS, Don		Man-Powered Aircraft	
		Tab Books Blue Ridge Simmit, PA	1979
WOODWARD, C.R.	**	Scientific Cycle Training	
		Temple Press, London	1960

TECHNICAL REPORTS

LARRABEE, Gene	**	Minimum Induced Loss Propellers	1983
N.A.C.A	**	Aerodynamic Characteristics of Airfoils	1928
		(Nat.Advisory Committee on Aeronautics) TL574 A4.U52	

CATALOGS

SHIMANO	**	Small Bicycle Parts	1993

Best reference sources: * Included in whole herein; ** Relevant material included herein.

Other Good
RELATED BOOKS

These are some of the very best, and are recommended reading.

Angelucci, Enzo, and Paolo Matricardi, *World Aircraft* (Chicago: Rand McNally, 1979).
Bailey, B., *To Fly Like a Bird* (Folkestone, U.K.: Bailey Brothers & Swinfen, 1976).
Barringer, Lewin B., *Flight Without Power* (New York: Pitman Publishing, 1940).
Burke, James D., *The Gossamers: A Case Study in Design*
    (Pasadena, CA: Aerovironment, 1980).
Dalton, Stephen, *The Miracle of Flight* (New York: McGraw-Hill, 1977).
DeDera, Don, *Hang Gliding: The Flyingest Flying*
    (Flagstaff, Arizona: Northland Press, 1975).
Dwiggins, Don, *Man-Powered Aircraft* (Blue Ridge Summit, PA: Tab Books, 1979).
Foote, Timothy, ed., *The World of Bruegel* (New York: Time-Life, 1968).
Gibbs-Smith, Charles H., *Aviation: A Historical Survey*
    (London: Her Majesty's Stat. Office, 1970).
Gilbert, James, *The Flier's World* (New York: Grosset & Dunlap, 1976)
Green, William, ed., *The Aircraft of the World* (London: Macdonald, 1965).
Gunston, Bill, *The Complete Story of Man's Conquest of the Air*
    (London: Octopus Books, 1978).
Guttery, T. E., *The Shuttleworth Collection* (Biggleswade, U.K., 1976).
Jameson, William, *The Wandering Albatross* (New York: William Morrow, 1959).
Kimura, Hidemasa, *Man Powered Aircraft Since 1963* (Tokyo: Nihon Univ., 1977).
Poynter, Dan, *Hang Gliding* (Santa Barbara, CA: Parachuting Publications, 1977).
Reay, David A., *The History of Man Powered Flight* (Oxford: Pergamon Press, 1977).
Riedel, Peter, *Start in den Wind* (Stuttgart: Motorbuch Verlag, 1928).
Robins, James G., *The Wooden Wonder* (Edinburgh: John Bartholomew, 1974).
Ruppell, Georg, *Bird Flight* (New York: Van Nostrand Reinhold, 1977).
Schmidt-Nielsen, Knut, *How Animals Work* (London: Cambridge Univ. Press, 1972).
Schulze, Hans-Georg, and Willi Stiasny, *Flug Durch Muskelkraft*
    (F.a.M: Verlag Fritz Knapp, 1936).
Sherwin, Keith, *Man Powered Flight*
    (Watford, U.K.: Model & Allied Publications, 1975).
Simons, Martin, *Model Aircraft Aerodynamics*
    (Watford, U.K.: Model & Allied Publications, 1978).
Taylor, John. H., eds., *The Guinness Book of Air Facts*
    (Enfield, U.K.: Guinness Limited, 1973).

Airship
ORGANIZATIONS

AIRSHIP COMMITTEE	*Amer. Institute of Aero. & Astro.*	703 / 264-7500
Win Arata, Chmn.	1801 Alex. Bell Drive #500	fax: 264-7551
(805 / 937-9044	Washington, DC 201	

The *Airship Technical Committee* was originally part of the Lighter-Than-Air Unit since 1987. As of 1997, because of the specialized interest, it focused its endeavors on promoting and disseminating technical information to those with a similar interest. The ATC meets twice a year at different ends of the country (usually May & Oct.). They meet as a group on odd number years (mid Summer), sponsoring an International Airship Convention. Activities are reported in "Airship" and "Gas Bag" magazines.
Web Site: http://www.aiaa.org

AIRSHIP COMMITTEE,	*Balloon Federation of America*	515/961-8809
Rick Wallace, Chmn.	P.O. Box 400	fax: 961-3537
(310 / 451-4233)	Indianola, IA 50125	bfaoffice@aol.com

The *Airship Committee* formed as an active group in 1987 to promote the development of Thermal Airships as a competitive sport. As part of the BFA, It is authorized by the Federacion Aeronautique Internationale (FAI) to officiate world records, and to hold periodic events to select U.S. representatives to World Championships. As the membership is very small, they usually meet informally at major LTA events. No Web Site at this time. http:www.bfa.ycg.org

AIRSHIP ASSOCIATION	*Royal Aeronautical Society*	44-1303-277650
Arnold Naylor, Chmn.	4 Hamilton Place	fax: 1303-277650
(44-01-499-3515)	London, W1V 0VQ, U.K.	naylor@airship.demon.co.uk

The Association was formed within the RAS to promote and disseminate technical information to those with an interest in Airships. The Technical Committee meets periodically at the downtown Royal Aero Club building in London. All activities are reported in their bimonthly magazine "Airship". They meet as a group on even number years (mid Summer), sponsoring an International Airship Convention. Activities are reported in "Airship" and "Gas Bag" magazines.
Web Site: http://www.airship.demon.co.uk

AIRSHIP GROUP	*British Balloon & Airship Club*

The BBAC is a very active organization for the promotion of sporting and social aspects of LTA. Airship endeavors are mainly in Thermal Blimps, but all methods of LTA flying are held in high regard. They report all airship activities in their monthly publication "Balloons & Airships".

Conclusion

Long before the sun melted Icarus's wings, man has longed to fly under his own power, but even Leonardo da Vinci was forced to give up this quest after 16 years. Faced with some irrefutable facts of physics, man has all but ceased trying to imitate the bird. Instead, he has turned his attention to a more favorable part of the anatomy: the legs.

Ever since Henry Kremer, an industrialist, announced a sizable cash prize for the person who piloted the first pedal-driven plane in a figure-eight, under the stipulations laid down by the Royal Aeronautical Society, man has been pedaling for the stars; and, as the accompanying information suggests, he has lost little interest in pedal-powered flight.

While there is a fair chance man-powered flight might evolve as a sport and, as enthusiasts hope, an Olympic event, it is likely that pedal power can be employed most productively on planet Earth. Reports from many developing countries indicate that, with the prohibitive cost of fuel oil, pedal power is re-emerging as an important factor in transportation and work. The fact is that, under many circumstances, pedal power is as "appropriate" and effective as the low-horsepower internal combustion engine. And, as the bicycle in a sense "liberated" people at the turn of the century, pedal power can liberate millions again. Surely pedal power has the potential for humanizing labor.

Pedal power is different things to different people. Maybe the most useful service it can perform in Third World Countries to sustain the lives of those with marginal resources. On the other hand, Americans might be more interested in the sporting aspects.

But perhaps the result is the same. If the numerous applications of pedal power make us less dependent on finite resources, then we satisfy a common objective. And if pedal power extends beyond class and economic lines, we have put geography to rest.

In some respects pedal power is a fitting symbol of appropriate technology: It can be applied in numerous ways to satisfy a variety of conditions. Yet pedal power, as the subject of serious research, is in the infant stage, where inventions appear to feed the field "by the day."

For that reason, the editor (and contributors) would greatly appreciate ideas and suggestions from readers pertaining to developing technology, and new discoveries in this exciting area. Correspondence on this subject would be greatly appreciated by:

James C. McCullagh
c/o RONDALE PRESS
Emmaus, PA

"We know it can be done, but better and more efficiently"

ISBN LOG BOOK

OTHER Publications by Robert J. Recks

ISBN	Title	Date
0-937568 - 00 - 7	PROBLEMS OF HELICOPTER USE	6/1961
0-937568 - 01 - 5	HELICOPTER WEIGHT AND BALANCE	8/1961
0-937568 - 02 - 3	BALLOONING EXAMINATION GUIDE	9/1969
0-937568 - 03 - 1	BALLOONING TRAINING MANUAL	10/ 69
0-937568 - 04 - X	BALLOONING CLUB INTRODUCTION	11/69
0-937568 - 05 - 8	BALLOONING FLIGHT CURRICULUM	12/ 69
0-937568 - 06 - 6	BALLOONING PROPOSAL, FAR-AMENDMENTS	1/1970
0-937568 - 07 - 4	BALLOON COMPETITION & SANCTION GUIDE	2/1970
0-937568 - 08 - 2	BALLOON PILOT SKILL RATING	4/1970
0-937568 - 09 - 0	TEACHING BALLOONING TO ADULTS	5/1970
0-937568 - 10 - 4	BRIEFING BALLOON CREW MEMBERS	6/1970
0-937568 - 11 - 2	BALLOONS & PILOTS CURRENTLY ACTIVE	7/1970
0-937568 - 12 - 0	BALLOON RECORDS, NATIONAL & INTERNATIONAL	8/1970
0-937568 - 13 - 9	KITE BALLOONING, HISTORICAL NOTES	1/1971
0-937568 - 14 - 7	FREE BALLOONING, HISTORICAL NOTES	2/1971
0-937568 - 15 - 5	BUILDING YOUR OWN SPORT BALLOON, VOL-1	1/1975
0-937568 - 16 - 3	BUILDING YOUR OWN SPORT BALLOON, VOL-2	8/1981
0-937568 - 17 - 1	BUILDING YOUR OWN SPORT BALLOON, VOL-3	3/1975
0-937568 - 18 - X	BALLOONS ON STAMPS, 1783-1983	12 / 80
0-937568 - 19 - 8	HELICOPTER EXTERNAL LOSS, VOL-1	3/1981
0 937568 - 20 - I	AIRSHIPS ON STAMPS	6/1981
0-937568 - 21 - X	HELICOPTER EXTERNAL LOADS, VOL-II	8/1981
0-937568 - 22 - 8	AIRSHIPS & BALLOONS ON STAMPS	10 / 81
0-937568 - 23 - 6	BALLOON STAMP HISTORY ALBUM	12 / 81
0 937568 - 24 - 4	IT BEATS WORKING, GYPSY PILOT AUTOBIO	Res
0-937568 - 25 - 2	AIRCRAFT TERMS, ENGLISH-PORTUGUESE	12 / 82
0-937568 - 26 - 0	WHO'S WHO OF BALLOONING, 1783-1983	11 / 83
0-937568 - 27 - 9	BALLOONING WITH SMOKE	Res.
0-937568 - 28 - 7	BUILDING SMALL GAS BLIMPS	12 / 97
0-937568 - 29 - 5	BUILDING THERMAL BLIMPS	1/1998
0-937568 - 30 - 3	MUSCLE POWERED BLIMPS	9/1998

ABOUT THE AUTHOR

Born and raised in Southern California; Served as a mechanic in the U.S.Army; Engineering major at Compton College; Started flight training; Graduated from the Univ.of Miami (FL); Served as a flight instructor; Flew worldwide as an Airline Engineer/Pilot, Active Balloon & Airship pilot since the 1960's; Served as a FAA Designated Airworthiness Rep. specializing in Lighter-Than-Air aircraft; Received a Master Riggers License after 30 years in Parachuting; Serves as an Aviation Consultant specializing in Inflatable Technology.

www.ingramcontent.com/pod-product-compliance
Lightning Source LLC
Chambersburg PA
CBHW080008210526

45170CB00015B/1921